SpringerBriefs in Archaeology

Contributions from Africa

Series Editor
Ann B. Stahl
Department of Anthropology
University of Victoria
Victoria, British Columbia
Canada

More information about this series at http://www.springer.com/series/13523

Shadreck Chirikure

Metals in Past Societies

A Global Perspective on Indigenous African Metallurgy

 Springer

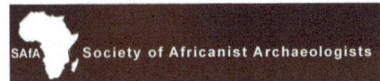 Society of Africanist Archaeologists

Shadreck Chirikure
Department of Archaeology
University of Cape Town
Cape Town
South Africa

ISSN 1861-6623 ISSN 2192-4910 (electronic)
SpringerBriefs in Archaeology
ISBN 978-3-319-11640-2 ISBN 978-3-319-11641-9 (eBook)
DOI 10.1007/978-3-319-11641-9

Library of Congress Control Number: 2014959261

Springer Cham Heidelberg New York Dordrecht London
© The Author 2015
This work is subject to copyright. All rights are reserved by the Publisher, whether the whole or part of the material is concerned, specifically the rights of translation, reprinting, reuse of illustrations, recitation, broadcasting, reproduction on microfilms or in any other physical way, and transmission or information storage and retrieval, electronic adaptation, computer software, or by similar or dissimilar methodology now known or hereafter developed.
The use of general descriptive names, registered names, trademarks, service marks, etc. in this publication does not imply, even in the absence of a specific statement, that such names are exempt from the relevant protective laws and regulations and therefore free for general use.
The publisher, the authors and the editors are safe to assume that the advice and information in this book are believed to be true and accurate at the date of publication. Neither the publisher nor the authors or the editors give a warranty, express or implied, with respect to the material contained herein or for any errors or omissions that may have been made.

Printed on acid-free paper

Springer is part of Springer Science+Business Media (www.springer.com)

For our sons Tawana and Tadana

Acknowledgements

When asked to be part of this important project, I was very surprised as much as I was excited. Unreservedly, I owe a big debt of gratitude to Professor Ann Stahl, the Series Editor, for the invitation and for very stimulating, inspirational, intelligent and resolute guidance throughout the development of this work and the publication process. *Ndinotenda* (thank you so much) Ann. Teresa Krause at Springer laboured tirelessly together with Ann and the Series Advisory body to ensure that this much needed series materialized.

This work benefited from a thorough review by two anonymous readers whose incisive comments strengthened the initial ideas and resulted in the production of a more effective and geographically balanced outcome. I would be remiss if I did not acknowledge the immortal contribution of the late Professor P. J. Ucko who always and in his own ways challenged me to articulate aspects of African archaeology to a global audience.

It is almost 15 years since I first knocked on Professor Thilo Rehren's door on the third Floor of the Institute of Archaeology at UCL. That day marked my first induction into the study of metals in archaeology. Since then, Thilo has been a consistent, very patient and untiring mentor in all technical aspects of archaeometallurgy, from the use of analytical equipment to interpreting the results. Even in the drafting of this book, Thilo as always, had many words of wisdom to share amidst his many other innumerable commitments. Andrew Reid made sure that the technical details that I learnt from Thilo were placed in their social context. Professors David Killick and Simon Hall took me under their wing when I had just graduated from the Institute of Archaeology in 2005. Since then, David gave me countless tutorials in determinative mineralogy and assisted me in assembling a reference collection of mineralogy books and mineral specimens. Simon allowed me to participate in his many fieldwork projects in and outside South Africa. David and Simon generously shared their unpublished work which helped me to navigate profitable research avenues. Dr. Duncan Miller, another longtime supporter, is always available to attend to my frequent requests and also gives me unlimited access to his vast and staggering experience. I therefore feel very lucky to have Simon Hall, David Killick, Duncan Miller, Andrew Reid and Thilo Rehren as faithful mentors, colleagues, critics and above all friends. To them I say, thank you immensely; without your generosity and goodwill, I would be very far away from where I am today. I must also acknowledge the many discussions that I had with Dr. Foreman Bandama and Ms. Abigail (Moff) Moffet in the Archaeological Materials Laboratory at the University of Cape Town which also gave this work its present character. Dr. Tim Maggs also contributed in many ways and generously gave me access to his vast archive and experience which made life a lot easier.

Along the way many colleagues supported me in various ways but I would like to single out Judith Sealy, Gilbert Pwiti, Bill Dewey, Robert Heimann, Chap Kusimba, David Reid, Mark Pollard, Munyaradzi Manyanga, Innocent Pikirayi, Jane Hubert, Andrew Reid, Phil De Barros, Per Ditlef Fredriksen, Godfrey Mahachi, Darlington Munyikwa, Thomas Thondhlana, K T Chipunza, Marcos Martinon-Torres, Dana Drake Rosenstein, Tom Fenn, Webber Ndoro, Randi Haaland, Susan McIntosh, Seke Katsamudanga, Joseph Chikumbirike, Mukundi Chifamba, Abigail Moffet, Bertram Mapunda, Mzee Edwinus Lyaya, Jane Humpris, Paul Sinclair, Martin Carver, Chris Scarre, Paul Lane, Rob Morrell, Pamela Eze-Uzomaka, Augustine Holl, Akin Ige, Ian Freestone, Adria La Violette, Mark Horton and others too numerous to mention. Although space precludes me from naming everybody by name, I sincerely appreciate and value the support provided by colleagues and critics alike.

The National Research Foundation of South Africa funded many of my research projects and provided grants which made the writing of part of this book possible. The NRF, through an Indigenous Knowledge Systems Grant, partly funded my sabbatical at Harvard and as such, I am extremely grateful for their support. The University of Cape Town Research Committee and the University of Cape Town Research Office through the Africa Knowledge Project and the Programme for Enhancement of Research Capacity has also been a consistent funder. I thank Professor Danie Visser, the UCT Deputy Vice Chancellor for Research, Dr. Marilet Sienaert the Executive Director for Research at UCT together with Professor Rob Morrell, the coordinator of PERC. Special thanks also go to Professor Thandabantu Nhlapo for resolute support and encouragement. Additional funding from the UCT Faculty of Science Awards is acknowledged with sincere gratitude. I therefore thank Professor Anton Le Roex, the Dean of the Faculty of Science for his goodwill. The Institute for Archaeometallurgical Studies (IAMS), Institute of Archaeology, UCL is an unwavering supporter. I therefore sincerely thank the trustees of IAMS. The Wenner Gren Foundation played an important role in my early career development and continues to do so in various direct and indirect ways. At the Wenner Gren Foundation, I extend my appreciation to President Leslie Aiello, Judy Kreid, Mike Muse and their colleagues for the sterling effort in promoting all types of anthropological research.

I also acknowledge the trustees of the Mandela-Harvard Fellowship for making my stay at Harvard possible. Some of the draft chapters for this book were written during my time as a Mandela-Mellon Fellow at Harvard University's Hutchins Centre for African and African American Studies in 2012. At the Hutchins Centre, I would like to thank Professor Louis 'Skip' Gates, Krishna Lewis, Abi Wolf and Donald Yacovone. Emmanuel Akeyampong, Jean Comaroff, Suzanne Blier, Rowan Flad and Ben Lewis also shared with me their knowledge of various issues that made my time at Harvard productive. While at Harvard, Charles van Onselen and my fellow fellows Celia Cussen and David Bindman, were fantastic sparring partners.

Dr. Pamela Smith kindly granted permission to publish the roped bronze object from Igbo Ukwu from the late pioneer of African archaeology, Professor Thurstan Shaw's estate. Thank you immensely, Dr. Smith, for your support to Thurstan's vision and interest in a fully-fledged African archaeology. The Barbier-Mueller Gold of Africa Museum in Cape Town granted permission to publish outstanding images of masterpieces in African gold working. Professor Phil de Barros another longtime colleague generously availed photographs from West Africa. Dr. Jane Humpris also granted me permission to publish her Meroe images while Dr. Tim Maggs provided the image of the twin furnaces from kwaZulu-Natal. Dr. Foreman Bandama drew all the maps and other line drawings (except where acknowledged) in the book at a time when he had to juggle many other important things. *Ndinotenda babamudiki*—thank for your support.

On so many occasions, the writing of this book took me away from my family and friends. To them I say thank you immensely for your unwavering support. May God bless all of you!

I would be remiss if I do not mention the immense contribution which my best friend and companion Geraldine always makes in my life. To you my wife, thank you so much for taking care of everything when I was busy with this project.

While these individuals and institutions assisted, errors and omissions that remain should solely be attributed to me.

Contents

1 **Metals and the Production and Reproduction of Society** 1
   Introduction 1
   Data Sources 3
   Towards an Integrated View of Global Metallurgy 10
   Organization of Work 12
   References 13

2 **Origins and Development of Africa's Preindustrial Mining and Metallurgy** 17
   Introduction 17
   Origins of Metallurgy in Egypt and Adjacent Areas 19
   Metal from Somewhere: On the Origins of Metallurgy in West, Central and East Africa 20
   Why are Iron and Copper Earlier than Gold and Bronze in Sub-Saharan Africa? Some Provocative Thoughts 28
   Conclusion 30
   References 31

3 **Mother Earth Provides: Mining and Crossing the Boundary Between Nature and Culture** 35
   Introduction 35
   History of Mining: A Global Perspective 38
   The First Step: Ore Exploration and Prospecting in Precolonial Africa 39
   Methods of Crossing the Nature–Culture Boundary: Mining of Ores in Preindustrial Africa 40
   Surface Collection 40
   Alluvial Mining 41
   Open Mining 44
   Underground Mining 45
   Preindustrial Hoisting and Beneficiation 51
   Mining Equipment and Other Paraphernalia 53

xi

	The Anthropology of Mining: A Global Outlook	53
	Conclusion	56
	References	57
4	**Domesticating Nature**	61
	Introduction: Transforming Ore into Metal	61
	Raw Materials	63
	Brief Overviews: Metal Smelting in Preindustrial Africa—	
	Egypt, Nubia, North Africa and the Horn of Africa	67
	West Africa	74
	Central Africa	78
	East Africa	82
	Southern Africa	84
	Anthropology of Smelting	87
	Conclusion	92
	References	93
5	**Socializing Metals**	99
	Introduction: Fabricating Metal into Cultural Products	99
	Metal Fabrication in Egypt and Nubia	101
	Forging, Smithing and Casting in Sub-Saharan Africa: West and	
	Central Africa	105
	East and Southern Africa with Occasional References	
	to Sudan and Ethiopia	111
	Metals in Society: The Anthropology of Smithing and Metal Objects	119
	Conclusion	121
	References	122
6	**The Social Role of Metals**	125
	Introduction	125
	General Impact of Metals in Society	126
	Metals, Sociopolitical Complexity and Urbanism	130
	Metallurgy, Culture Contact (Interaction), Proto-Globalization	
	and Technology Transfer	138
	Conclusion	145
	References	146
7	**Bridging Conceptual Boundaries, A Global Perspective**	151
	Introduction	151
	African Metallurgy and the Bridging of Conceptual Boundaries	
	Between Technology, Society and Culture	153
	Bridging Analytical Boundaries: From Sources of Ethnographies	
	to Domains of Integrated Studies	154
	Local Responses to Technology Transfer and Knowledge Dispersal	157

African Metals: Land and Sea Links and Protoforms of Globalization 158
Changing Contexts of Knowledge Production and the Future of
African Preindustrial Metallurgy 159
Conclusion 161
References... 162

Index.. 165

About the Author

Shadreck Chirikure graduated with a Master of Arts in artefact studies degree and a PhD in archaeology from the Institute of Archaeology, University College London. Chirikure was born into one of the most senior houses of the Gutu-Rufura people in rural Zimbabwe. During his childhood years, Chirikure's grandmother was a potter and a number of men forged scrap iron in his village. Therefore, negotiating through rituals and taboos embedded in pottery making and other categories of practice were part of Chirikure's experience growing up. Unknown to him, these would be part of his academic routine when he later became an archaeologist. Because of this village experience, Chirikure always attempts to gestate archaeological reconstructions that are tempered with local realities where nothing was fixed in space and time. His main research marries techniques from earth and engineering sciences with those from more humanistic disciplines to study high temperature technologies such as metallurgy and pottery making to enlighten their contribution to societal development. Currently, Chirikure's work on the metallurgy of the world heritage sites of Great Zimbabwe and Khami in Southern Africa is throwing new light on the contribution of metals to culture contact, interaction and social differentiation. The work shows that metals, like cattle were a pivot on which society achieved growth and renewal. Shadreck has published extensively on the subject including a book, multiple journal articles, book chapters and contributions to prestigious encyclopaedias. In the process, he won several national and international awards for his contributions to African Iron Age research.

List of Figures

Fig. 1.1	Links between metal production and use with various dimensions of society	2
Fig. 1.2	Selected well-known sites and metal working groups by region. Sudan, Ethiopia and Eritrea are clustered under North Africa because they share the same metallurgical transition from copper through bronze to iron	4
Fig. 1.3	Photomicrograph of iron-smelting slag from Mapungubwe Hill, Southern Africa showing a magnetite skin near top right corner, egg-shaped wüstite, skeletal fayalite on a glass matrix and some voids in dark black. Magnetite skins form when slag from furnaces is exposed to cool air and is indicative of smelting and in some cases slag tapping	8
Fig. 1.4	Integrated view of preindustrial metal production and use outlining the inputs and outputs at various stages and the accompanying anthropological factors	11
Fig. 2.1	Map of Africa showing metalworking sites with some of the most important highlighted by number	18
Fig. 2.2	Ellingham diagram shows the ease with which metals and sulphides can be reduced. The position of the line for a given reaction on the Ellingham diagram shows the stability of the oxide as a function of temperature. Reactions closer to the *top* of the diagram are the most 'noble' metals (for example, gold and platinum), and their oxides are unstable and easily reduced. Moving towards the *bottom* of the diagram, metals become progressively more reactive and their oxides harder to reduce	24
Fig. 3.1	Location of mines worked in African antiquity. Note that because of its abundance, iron was worked in more areas than illustrated here	36
Fig. 3.2	Akan gold miners diving into the Ankobra River to extract diamond-rich sand which was panned on the river bank.	42

Fig. 3.3	Cross section of the Aboyne gold mine, Central Zimbabwe	47
Fig. 3.4	Umkondo copper mine in Zimbabwe where miners strategically backfilled and opened up new shafts to create pillars (unmined blocks) for structural stability.	48
Fig. 3.5	Wooden bucket found inside ancient gold mine in Southwestern Zimbabwe and donated to Natural History Museum, Zimbabwe. Exact provenance unknown.	51
Fig. 3.6	Porcupine quills used for storing gold.	51
Fig. 3.7	Undated grinding stone with dolly holes used to crush copper ores at Phalaborwa.	52
Fig. 4.1	Known metal smelting groups in Africa. *Numbers* indicate key sites and landscapes	62
Fig. 4.2	Approximate distribution of bowl, shaft and natural draught furnace types in Africa.	66
Fig. 4.3	Distribution of bellows across sub-Saharan Africa.	68
Fig. 4.4	Location of early Egyptian and Nubian smelting sites.	69
Fig. 4.5	Location of sites with smelting evidence in North Africa, Egypt, Nubia and Ethiopia	70
Fig. 4.6	An Old Kingdom pictorial showing six smelters blowing into two crucibles (Mastaba of Mereruka, fifth Dynasty)	71
Fig. 4.7	Bowl and shaft furnaces used in dynastic Egypt	72
Fig. 4.8	Furnace F5 excavated by Shinnie (1985) at Meroe.	73
Fig. 4.9	Large iron slag mound at Meroe.	73
Fig. 4.10	Smelting sites in West Africa	74
Fig. 4.11	Second-millennium AD furnaces used in Bassar, Togo.	75
Fig. 4.12	Second-millennium AD slag mounds at Bassar, Togo.	76
Fig. 4.13	Remnants of a circular shaft furnace from an iron smelting furnace in Senegal, dated between the twelveth and fourteenth centuries AD. The black material is slag that solidified within the furnace.	77
Fig. 4.14	Location of Central African sites	79
Fig. 4.15	Cross section of Mafa down draught furnace.	80
Fig. 4.16	Location of East African sites and groups	83
Fig. 4.17	Location of Southern African sites and groups	85
Fig. 4.18	Twin-bowl furnaces used in the Later Iron Age (c. AD 1700 onward) of KwaZulu-Natal, South Africa.	86
Fig. 4.19	Anthropomorphic low shaft iron smelting furnace from Nyanga, Eastern Zimbabwe. Note the molded breasts, navel and waist belt for enhancing fertility.	89
Fig. 4.20	Decorated furnaces, one depicting a woman giving birth. Redrawn from furnaces on display at Natural History Museum, Bulawayo	90
Fig. 4.21	Anthropomorphic drum, granary and iron smelting furnace from Bent (1896).	91

List of Figures xix

Fig. 5.1 Metal working groups and important sites discussed in the text 100
Fig. 5.2 Dish bellows connected to a clay nozzle with a reed 103
Fig. 5.3 Location of West and Central sites and groups discussed in
the text 106
Fig. 5.4 Traditional iron forging workshop at Bitchabe, Togo, using
concertina bellows 107
Fig. 5.5 Leaded tin bronze Igbo Isiah roped vessel produced using
the lost wax method 109
Fig. 5.6 Pectoral from the tumulus of Rao Senegal dating between
seventeenth and eighteenth centuries. 110
Fig. 5.7 Photograph showing earliest gold earing from Jenne Jeno,
Inland Niger Delta, Mali. 110
Fig. 5.8 Akan crocodile and lizard sword ornaments 111
Fig. 5.9 Location of smithing sites in East and Southern Africa 112
Fig. 5.10 Illustration of a Mambari smith by Holub (1881) showing
bellows leading to tuyeres and a very small forge. Note also
the tongs illustrated 113
Fig. 5.11 Illustration of pot bellows used for smithing by Mambari
smiths (Holub 1881) 113
Fig. 5.12 Goat skin bellows similar to the ones used by Njanja.
Approximate size 80 cm long. Natural History Museum,
Bulawayo, Zimbabwe 114
Fig. 5.13 X-shaped copper ingots in use in much of Southern
Zambezia in the Iron Age, particularly after AD1000 115
Fig. 5.14 Musuku ingot housed at Iziko Museums, Cape Town 115
Fig. 5.15 Half of a cross-shaped copper ingot mold (carved from
steatite?) used to produce ingots in 5.15 recovered from
Zimbabwe, on display at the Natural History Museum,
Bulawayo, Zimbabwe 117
Fig. 5.16 Second-millennium AD soapstone ingot mold with a gold
pellet insert, Natural History Museum, Bulawayo 118
Fig. 5.17 Femur and bangles from a high-status individual excavated
from the dry-stone-walled Zimbabwe culture site of
Danangombe (AD1680–1850), Central Zimbabwe. The
excavators retrieved this bone with those high numbers of
copper or copper alloy bangles, which are quite numerous.
It is on display in the Zimbabwe Museum of Human Sciences...... 118

Fig. 6.1 Location of metal working groups and places
discussed in the text 128
Fig. 6.2 Iron gong recovered from Great Zimbabwe. Natural History
Museum, Bulawayo 129
Fig. 6.3 Gold leaf bowl from Mapungubwe, Southern Africa. It is
believed that the leaf was attached to a wooden core which
has since decayed (Miller 2001). 129

Fig. 6.4	Mapungubwe golden rhino made of gold leaf (size, c. 10 cm). It symbolized the "majesty" of kingship............................	129
Fig. 6.5	Location of prominent urban centers in Africa	132
Fig. 6.6	Indian Ocean cowrie and sea shells made their appearance in Southern Africa from AD 700 onwards	133
Fig. 6.7	Ingots from burials excavated from the Upemba depression in the modern-day DRC.	135
Fig. 6.8	Iron projectiles used by the Buluba of the Katanga region of the Democratic Republic of Congo. Natural History Museum of Zimbabwe, Bulawayo..	137
Fig. 6.9	Diamond-shaped hoes produced by Phalaborwa smiths in Northern South Africa between c. AD 1600 to 1900. The hoes were used as currency..	139
Fig. 6.10	Location of Rooiberg in relation to capitals.....................................	140
Fig. 6.11	Ming Dynasty porcelain from Zimbabwe housed at Iziko Museum, Cape Town.	142
Fig. 6.12	Regional and trans-continental connections between Southern and Western Africa and the trans-Saharan and Indian Ocean worlds	143

List of Tables

Table 2.1	shows some of the earliest dates for the appearance of metallurgy in Africa. Calibrated using OxCal version 4.2.3 Bronk-Ramsey (2013) and IntCal13 (Reimer et al. 2013)	21
Table 3.1	Techniques and tools used in precolonial mining across Africa. Adapted and modified after Hammel et al. (2002, p. 51)	54
Table 4.1	Chronology of ancient Egypt and Nubia	69

Chapter 1
Metals and the Production and Reproduction of Society

Introduction

As with archaeology, the development of archaeometallurgy—the study of pre-industrial metal production, distribution and consumption—is richly varied, from place to place and time to time (Childs and Killick 1993; Chirikure 2014; Craddock 1995; Herbert 1993; Killick and Fenn 2012; Kiriama 1987; Kusimba 1993; Morton and Wingrove 1969; Okafor 1993; Rehren and Pernicka 2008; Rothenberg 1970; Summers 1969). Traditionally, archaeometallurgy was preoccupied with exploring the materiality of ore to metal and metal to artefact transformations from combined technological, economic and environmental perspectives. It was less concerned with the sociocultural dimensions of those materialities and processes (Hauptmann 2007; Joosten et al. 1998; Morton and Wingrove 1969; Rehren et al. 2007). However, it is self-evident that archaeometallurgy can no longer afford to ignore the anthropological dimensions which constituted a resilient component of preindustrial metal production technological repertoires and styles (Childs 1991; Cline 1937; Lechtman 1997; Rehren et al. 2007; Rickard 1939). Therefore, it is universally acknowledged that metal production and use in the past was simultaneously technical and sociocultural, with the result that it produced and reproduced society. This makes it highly problematic to study materials without considering the active role of 'materialities' in social life (Childs and Herbert 2005; Childs and Killick 1993).

Modern archaeometallurgical work continually demonstrates that as an integral component of human society and culture, the advent and subsequent entrenchment of metallurgy enabled humanity to respond to various necessities and fuelled appetites for luxuries (Fig. 1.1). In the process, the production and consumption of metals gestated sociopolitical inequalities which strategically positioned individuals and collectives on different socioeconomic and political gradients (Chirikure 2007; Chirikure and Rehren 2006; de Barros 1988; Hauptmann 2007; Miller 2002; Pollard and Bray 2007; Rothenberg 1999; Schmidt 1997; Srinivasan 1994). The production and use of metals also connected localities, regions and continents for millennia, initiating 'proto-globalization'. For example, trade in African gold played a pivotal role in the rise and fall of centres of power, trade and empires on

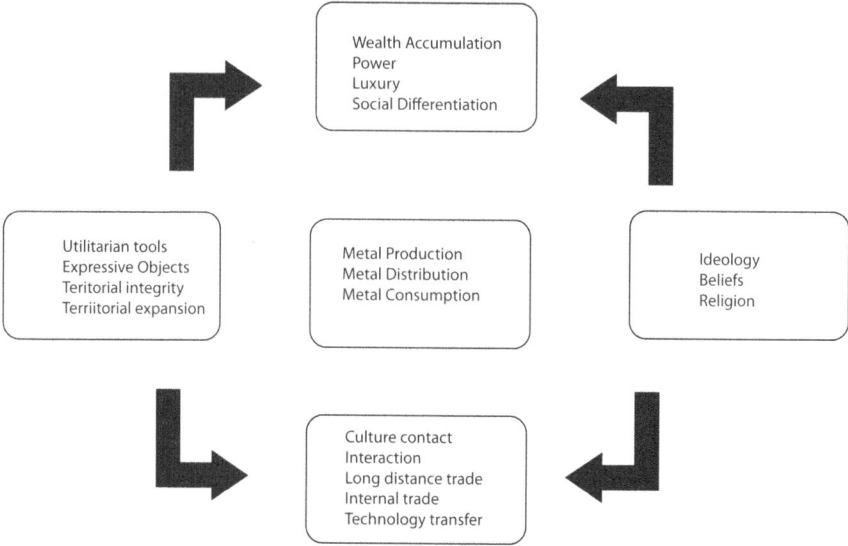

Fig. 1.1 Links between metal production and use with various dimensions of society

the continent throughout ancient times (Klemm and Klemm 2012). By 2400BC, Egyptians were trading gold from the Eastern Desert with Mesopotamia, generating immense wealth which was invested in monumental architecture and objects of conspicuous consumption (Garrard 2011). Similarly, West and Southern African gold financed developments in the Islamic world and adjacent territories from the mid-first millennium AD onwards (Levtzion 1973; Phimister 1974). This initiated cultural, technological and values exchange which, at different temporal scales, affected directly and indirectly the societies participating in this trade. The mining, production, consumption and distribution of different metals created new dynamics of power and interaction, which in turn shaped the evolution of society at various levels.

Due to a very high number of researchers per capita in some regions compared to others, our knowledge of the quartet of mining, smelting, smithing and distribution is uneven (Chirikure 2014). Rather than being a constraint, this differential rate of disciplinary emphasis and growth is a massive opportunity for reflection and inspiration because of the varied research trajectories in different world areas. For example, a brief navigation of the landscape of studies of preindustrial metallurgy illustrates vividly that we know considerably more about the origins of metallurgy in some parts of Eurasia than elsewhere in the world (Pringle 2009; Radivojević et al. 2010; Roberts 2009; Zangato and Holl 2010). While our grasp of the beginnings of metallurgy in Africa is mostly shrouded in mists of uncertainty, the anthropology of metal production and consumption on the continent leads the way (Childs and Herbert 2005; Schmidt 2009). Comparatively, the study of the technology of metal production, distribution and consumption in Europe is reasonably well known, particularly because of interest shown among chemical, earth and engineering sciences (Gale and Stos-Gale 1982; Morton and Wingrove 1969; Pernicka et al. 1997; Pollard et al. 2011).

Despite this uneven emphasis, Childs and Herbert (2005) and Killick (2014a) contend that studies of sub-Saharan Africa's preindustrial metallurgy are well ahead of those of other world areas including Egypt, Nubia, Ethiopia and North Africa. Therefore, as varied as the contours of knowledge between the world's geographical regions might be, Africa's long but richly varied history of encounters with both cultural and technological aspects of metallurgy will help global archaeology to catch up in this respect. This provides the motivation for this work: with well-established cadastral points, such as the stages in the production cycle and the synergy between technological and anthropological factors, this volume aims to articulate and contour salient features of preindustrial African metallurgy onto the global map (Fig. 1.2).

Data Sources

Given Africa's vast extent and the very long and remarkably variegated history of metal production and use, no source can, on its own, provide a credible story of the role of metals in past societies. However, a number of sources are available which, when considered in combination, provide an intellectually nourishing picture of Africa's preindustrial metallurgy and its multiple lessons for global archaeology. These sources range from archaeology and archaeometallurgy to art history, history (including historical linguistics) and ethnography. Traditionally, specialists from these individual disciplines work within closed boundaries, resulting in disciplinary isolation and compartmentalized knowledge of the same phenomena. With a view to developing synergy between various sources and disciplines, this volume draws on diverse perspectives, as outlined below.

Historical Sources: Documentary and Oral History Written records are an important archive of information on preindustrial African metallurgy. Their utility and availability vary from time to time and area to area. For example, while literacy has a 5000-year history in Egypt and a 4000 year one in Nubia (Edwards 2004), written records for regions such as Southern Africa only started in the last 1000 years (Summers 1969). In Egypt and Nubia, there exist some documents and wall paintings that describe metalworking deep in antiquity (Davies 1943). Pieterse (1998) describes Egyptian wall paintings that depict black Africans involved in copper working around 2500 BC. Metallurgical activities including the gold trade between Egypt and Mesopotamia are also captured in the Bible (Emery 1963; Scheel 1989). For other regions, written records appear sporadically from the time of Islamic contact around 700 AD when Islamic scholars wrote about West Africa (Levtzion and Hopkins 2000). For East Africa, there are occasional documents such as the Periplus of the Erythrean Sea (written in the first century AD)

Fig. 1.2 Selected well-known sites and metal working groups by region. Sudan, Ethiopia and Eritrea are clustered under North Africa because they share the same metallurgical transition from copper through bronze to iron

which discusses interaction between Greco-Romans and residents of towns such as Rhapta located on the Indian Ocean coast (Horton and Middleton 2000). By the early second millennium AD, Islamic chroniclers such as Ibn Battuta and Al Masudi provided detailed descriptions of events and places in West and East Africa. From the fifteenth century AD, written records appeared with increasing frequency as the Portuguese sailed the Atlantic and Indian Ocean littoral and set up forts and trading stations (Beach 1980; Mudenge 1974). Because they settled in areas such as Elmina in modern Ghana, and Sofala in Mozambique, the Portuguese left accounts containing various types of information about trade, politics and other aspects of life (DeCorse 1992), though these reports must be treated with caution because of inherent biases on the part of observers (Beach 1980).

Throughout sub-Saharan Africa, documentary records became more abundant during the last 300 years leading up to colonization (seventeenth, eighteenth and nineteenth centuries AD), some explicitly commenting on metallurgy. Mungo Park, David Livingstone and many other explorers and missionaries left reports which now form a critical archive of data on the various stages in the metal production cycle and their relations with other aspects of society (Cline 1937). The advent of industrialization as a form of organizing production in Europe promoted an early disappearance of preindustrial metal working processes there when compared to places such as Africa and Asia (Craddock 1995). However, despite the richness of the European historical record, which sometimes vividly captured metalworking events (see, e.g., *Georgius Agricola's De Re Metallica,* Hoover and Hoover 1950), few studies have drawn insight from history, myths and legends to develop a broad-based and informative understanding of Europe's precapitalist metallurgy (for exceptions, see Haaland 2004a; Martinon-Torres and Rehren 2009a). In India, attempts are being made to engage with written sources, but not on a scale comparable to Africa (see Tripathi 2013).

For the more recent periods in sub-Saharan Africa, oral history too is an important source of information for the study of preindustrial African metallurgy, particularly in areas with no written records. Oral history encompasses traditions, myths and legends and is transmitted by word of mouth from one generation to the next (Vansina 1985). In some areas of West Africa, griots or court historians passed on traditions for a very long time such that testimonies may go back to the first millennium AD (Levtzion 1973). Schmidt (1978) used oral historical accounts to develop a long-term perspective on Buhaya iron smelting, extending from the recent past to the early first millennium AD. In Southern Africa, historians such as Phimister (1974) used oral historical data to explore gold production in second millennium AD Southern Zambezia and its role in early state formation.

Like any other source, oral and documentary sources must be used with caution for they are often affected by a number of weaknesses, some source specific but others universally applicable. For example, oral traditions are easily forgotten, can be chronologically telescoped and may be edited to suit prevailing contexts (Beach 1980; Vansina 1985). Written records are also affected by observer bias, e.g. most Portuguese documents on the Shona of Northern Zimbabwe which describe silver mines that never existed (Mudenge 1988). Nowadays, there is a strong feedback problem whereby oral sources are affected by written history. With increasing literacy, African people read about their past, knowledge of which may filter into memory. Effectively, oral sources became documentary ones, and what was written became part of oral histories in a maze that is difficult to disentangle (Beach 1980). Fortunately, a large number of traditions recorded in the twentieth century are still available and can be used to cross check details. Whatever their limitations, historical sources often provide important details about events and historical processes of varying duration, something that archaeology struggles to address because time aggregates promoted by techniques such as radiocarbon dating compress events of different duration (Chirikure et al. 2012). Thus, when oral and documentary sources are combined with archaeology, essential complementary information is generated (Schmidt 1978; Stahl 1994).

Ethnographies Preindustrial methods of metal production and use in some areas of Africa ended only after industrialization introduced at conquest and in others persisted into the 1980s and 1990s (David et al. 1989; Huysecom and Augustoni 1997; Van der Merwe and Avery 1987). Thus, ethnographies are a rich source of technological and cultural details of metals and their role in society (David 2012). For example, ethnographic recording of the production chain of iron among the Dogon of Mali in West Africa has yielded a holistic understanding of technology in its social context (Huysecom and Augustoni 1997). In West-Central Africa, David et al. (1989) re-enacted Mafa iron smelting in the Mandara Mountains of Cameroon. The Mafa smelted magnetite sand in down-draught furnaces to produce a mix of cast iron and soft iron. Other ethnographic studies illuminated the symbolic and cultural attributes of preindustrial metal production and use among the Shona of Zimbabwe and Haya of Tanzania (Dewey 1991; Schmidt 1997). The value of these ethnographic studies lies in the fact that unlike the fragmentary archaeological record, they provide robust insight into technical and symbolic aspects of metal production and use. As such, ethnographies make it possible to explore how metal production was socially embedded. For example, through metal production, smelters reproduced and articulated ideas about fertility, witchcraft, power of ancestors, significance of medicines and pollution which fundamentally were rooted in societies that produced the metal (see Bent 1896 and Chap. 4).

There is a misperception that ethnographies are only useful in studying the preindustrial metallurgy of Africa. In India, significant amounts of ethnographic material were recorded over the centuries, but the application of ethnography to animate the archaeological record is still in its infancy (Keen 2013; Tripathi 2013). Similar observations can be extended to Latin America where, before and immediately after Columbus, indigenous methods of working silver were still widely present and documented (Cohen et al. 2010; Schultze et al. 2009). Indeed, ethnographies can also be applied to understanding Europe's preindustrial metallurgy. For example, Hansen (1986) makes it explicit that magic, myths and legends fundamentally lay at the heart of Medieval Europe's social and technological processes. In fact, Insoll (2008) has demonstrated that insight into ritual practice in Africa holds potential within limitations to inform on ritual practice in the European Bronze Age. Therefore, the rest of the world can learn from Africa, with its ethnography providing a source of information for understanding the metallurgical record.

Despite being useful, an uncritical application of ethnography is dangerous given the potential to extrapolate a presentist view onto the past. In continents such as Africa, the 'curse of the ethnography' is often criticized for creating a picture of a static continent devoid of any history and dynamism (Lane 2005). However, because some cultural principles are resilient, a careful application of ethnography supported by good contextual, spatial and temporal control has demonstrated the utility of combining ethnographies with archaeological data to distil more nuanced interpretations, particularly of processes that left little in terms of tangible remains (David and Kramer 2001; Stahl 1994, 2001). Therefore, the value of comparative insight drawn from historical and ethnographic sources, within well-defined contexts, can be a pivot on which researchers can base their comparison of different metallurgies and their impact on society across space and time.

Archaeology In spite of dealing with a fragmentary record, and lacking the finer-grained temporal resolution provided by ethnographies and histories, archaeology is essential for understanding metal production in past societies. Archaeology provides a long-term perspective critical for understanding the development of metal working through space and time. Furthermore, it provides context-specific examples of how various peoples and regions engaged with the world around them to produce and use metals. For example, Pleiner (2000) has on the basis of archaeology presented a diachronic picture of the evolution of iron production in Central Europe during the Iron Age. Pleiner observed changes in furnace types from slag pit furnaces characteristic of the Early Iron Age to slag taping furnaces typical of the Late Iron Age. In West Africa, Okafor (1993) noted significant changes in furnace operation between the Early and Late Iron Ages of Nsukka in Nigeria. While Nsukka Late Iron Age (AD1000–1600) furnaces were slag tapping, their predecessors in the Early Iron Age (600BC to AD1000) were non-slag tapping, thereby providing a picture of technological change over time. So too have differentials in the quantity of slag found at Early Iron Age sites when compared to those of the Late Iron Age in Southern and West Africa proffered insights into scale of production and evolution of specialization in the regions (Chirikure 2007; de Barros 2013). Archaeology also generates samples that can be studied using archaeometallurgical techniques and methods.

Archaeometallurgy Archaeometallurgy is an inter-disciplinary method for studying remains from past metal production and use (Rehren and Pernicka 2008). Because they are products of high-temperature processes, remains from preindustrial metalworking such as slags contain within their microstructures and composition partial histories of the processes which they have undergone (Bachmann 1882). Archaeometallurgists often deploy techniques from chemical, earth and engineering sciences to read important technical details such as the temperatures achieved in furnaces, the quality of ores and skills in manipulating redox conditions. Furthermore, by adopting more anthropologically oriented approaches that consider metallurgy to be simultaneously social and technical, it is possible to situate the 'human factor' in metalworking by looking at the decisions that smelters and smiths made for the process to be a success (Rehren et al. 2007). Even more pertinent is the observation that different stages in the *chaîne opératoire* of metalworking (Childs and Herbert 2005) left signatures on the landscape. When the material and cultural properties of process remains are considered in relation to the site's geography and surrounding landscape, a holistic view of technology in society is achieved. Finished metal objects too are significant for they also can be interrogated to provide information on provenance, trade and exchange as well as metal consumption (Fenn et al. 2009).

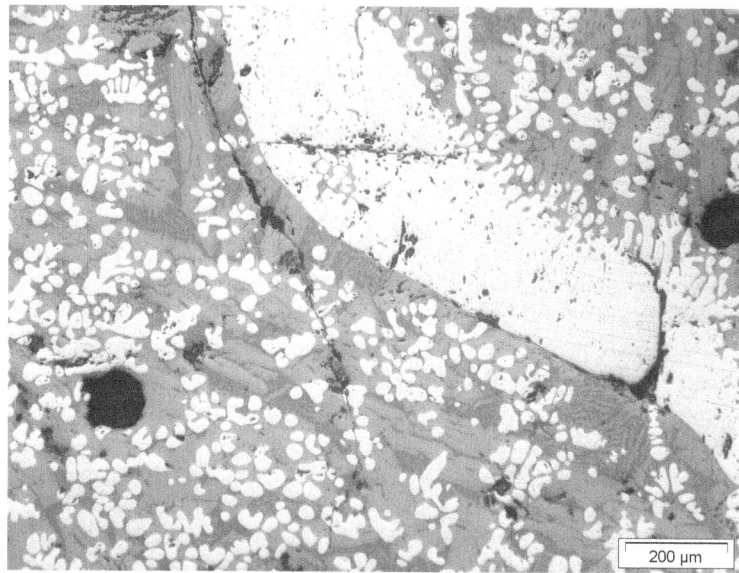

Fig. 1.3 Photomicrograph of iron-smelting slag from Mapungubwe Hill, Southern Africa showing a magnetite skin near top right corner, egg-shaped wüstite, skeletal fayalite on a glass matrix and some voids in dark black. Magnetite skins form when slag from furnaces is exposed to cool air and is indicative of smelting and in some cases slag tapping. (Photo credit: Author)

A number of complementary methods have been developed to gain information from archaeometallurgical remains, but archaeological fieldwork forms an important pedestal for all interpretations. As such, archaeometallurgy requires high standards of data recovery in the field. Data from laboratory techniques only make sense when interpreted in relation to the context of recovery and associated milieu. According to Chirikure (2014), the significance of fieldwork in archaeometallurgy cannot be overemphasized and is underlined by the fact that no matter how sophisticated laboratory methods are, the information from such techniques is meaningless if the methods of field recovery are poor. Where possible, it is crucial for archaeologists to have an expert archaeometallurgist in the field (Killick 2014b). Projects such as the Mandara Archaeological Project that included archaeometalurgists in their interdisciplinary teams (David et al. 1989) are an example to follow.

Generally, laboratory archaeometallurgy thrives on a combination of microscopic and compositional techniques (Bachmann 1982). Optical microscopic methods are based on the principle that when exposed to light under a plane-polarizing microscope (in reflected or transmitted plane-polarized modes), different phases in a polished sample of a material reflect light differently (Deer et al. 1992; Phillips and Griffen 2004;). For instance, the various phase characteristics of metallurgical slags provide information on efficiency of reduction (Morton and Wingrove 1972), furnace-operating temperatures (Chirikure and Rehren 2006) and the skills of the extinct smelters in balancing ore-to-fuel ratios and in reducing more oxide to metal (Rehren et al. 2007) (Fig. 1.3).

While useful, the limitations of light microscopes in relation to magnification (maximum of 1000 X) and an inability to determine the chemical composition of various phases can be transcended by recourse to analytical techniques such as scanning electron microscopy with attached energy dispersive spectrometer (SEM-EDS). SEM-EDS combines high image resolution techniques with analysis of the characteristic of X-rays produced when the sample is bombarded with primary electrons. This marriage produces a very powerful analytical tool suitable for analyzing small regions of solid materials and detecting spatial variation in composition. For example, crucibles can be characterized by looking at the composition of the clay matrix, the crucible slag and metal droplets to identify the metals worked and to understand the various contributions of various inputs to the metallurgical process (Bayley and Rehren 2007). The refractoriness of the clay as determined by the quantities of heat-resistant minerals such as kaolin informs on the skill of metallurgists in selecting suitable raw materials (Martion-Torres et al. 2006). Although possessing these distinct advantages, compositional data from the SEM are often not of a resolution low enough to detect some essential minor and trace elements in archaeometallurgical materials.

This limitation is addressed by resorting to techniques such as X-ray florescence (XRF) analysis. XRF exploits the properties of X-rays to identify the major, minor and trace element composition of archaeometallurgical materials (Pollard et al. 2011, pp. 93–118). Such compositional data are important for reconstructing raw material sources, and estimating furnace operating temperatures and the contribution of different variables to slag formation. Another useful technique is X-ray diffraction (XRD) analysis. XRD uses X-rays of known wavelengths to determine the lattice spacing in crystalline structures, and in the process, it indirectly identifies chemical compounds (Pollard et al. 2011, pp. 93–118). Raman spectrometry too is becoming an important technique in archaeometallurgy. It exploits the phenomenon that as radiation passes through a transparent medium, a small proportion of the incident beam is scattered in all directions (Muralha et al. 2011). The difference in wave length between the incident and scattered radiation is a characteristic of the material responsible for scattering (Pollard et al. 2011, pp. 83–85). Neutron activation analysis (NAA) is a powerful multi-element nuclear technique that allows the determination of the concentration of a large number of elements in archaeometallurgical remains. In essence, NAA involves converting some atoms of the elements within a sample into artificial radioactive isotopes by irradiation with neutrons (Pollard et al. 2011, pp. 195–208). The radioactive isotopes formed then decay to form stable isotopes at a rate which depends on their half-life. Measurement of the decay allows the identification of the nature and concentration of the original elements in the sample. Solid samples can be analysed and whole samples can be irradiated. Inductively coupled plasma-mass spectrometry (ICP-MS) too is best suited for rapid-trace element-level elemental analysis in archaeometallurgical materials.

All these geochemistry techniques have been applied with varying degrees of success in reconstructing the raw materials used in ancient metal production across the world. Matching the trace elements in the metal or alloy and ore is a common tool of tracing the movement of metal and ores on the landscape through trade and exchange relationships (Chikwendu et al. 1989; Pernicka et al. 1997). However, all analytical techniques do not have the same capabilities and detection limits, making it standard practice to combine one or more techniques in a single study. For example, Chirikure et al. (2010a) combined bulk chemical analyses using WDXRF with trace element analysis by ICP-MS, individual phase characterization using electron microprobe analysis (EMPA) and SEM to investigate the technology of tin smelting utilized at Rooiberg in the Southern Waterberg of South Africa. In other studies, Fenn et al. (2009) used lead isotope analyses in conjunction with compositional techniques to demonstrate the existence of contact between West Africa and Roman North Africa.

In summary, although different sources of information are available for studying preindustrial metal production and its role in society, they must be combined to create balanced picture of the past. As such, researchers must cross disciplinary boundaries to exploit interpretive leads offered by other disciplines. Fortunately, for studies of preindustrial metallurgy, archaeometallurgy is one of the most inter-disciplinary methods, combining data and techniques from history, archaeology, earth and engineering sciences, anthropology and sociology (Killick and Fenn 2012; Rehren et al. 2007). Such a broad-based view allows synergy to be developed between scientific techniques and archaeological or historical approaches to explore the socially embedded nature of metals in society.

Towards an Integrated View of Global Metallurgy

Although anthropologists such as Mauss (1954) have long realized that technology is socially mediated, it took a very long time for Western science to accept that it is impossible and in fact unwise to separate technical and cultural aspects of any given technology (Appadurai 1986). The tendency in Western science in the twentieth century was to bifurcate technology into the scientific and magical aspects. Ironically, before the onset of Europe's scientific revolution, which began after the Middle Ages, magic formed a key component of Western worldview (Hansen 1986). Because of its association with symbolism, indigenous African metallurgy was often derided as magical and therefore unworthy of proper scientific study (Schmidt 1996). From the 1960s onwards, Africanists realized that the so-called magic was just as important as the control of air pumped into the furnace or the quality of the ore. The rituals and taboos were a technology of practice that enabled smelters to take control of the process through learned behavior (Herbert 1993). Not surprisingly, when studies of African metalworking flourished, a conscious attempt was made to combine both the scientific and symbolic aspects of the technology (Herbert 1993; Killick 1990; Schmidt 1978; Van der Merwe and Avery 1987) (Fig. 1.4).

While these developments were taking place in Africa, the tendency in Anglo-American technological studies was to isolate science from anthropology. It was only in the early 1990s with the emergence of conceptual approaches such as *chaîne opératoire* that it was realized that technology is made up of cultural and scientific

Towards an Integrated View of Global Metallurgy

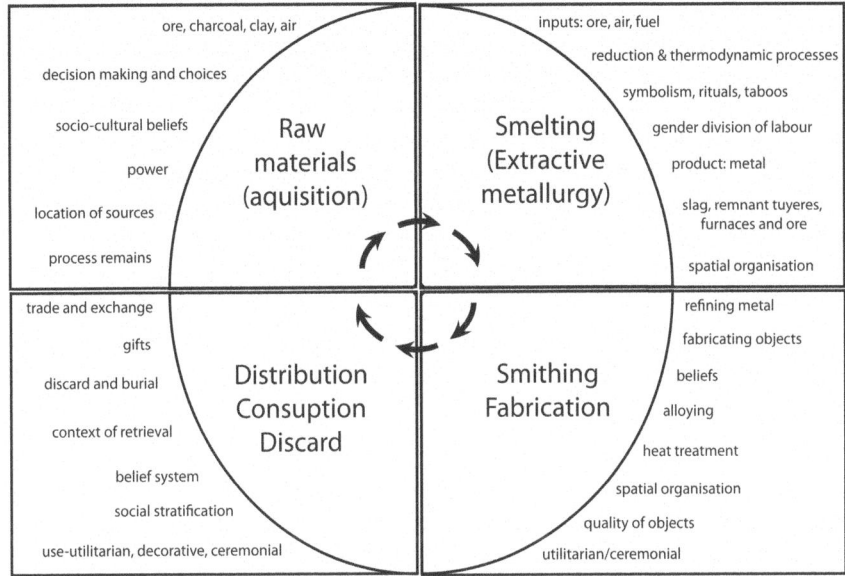

Fig. 1.4 Integrated view of preindustrial metal production and use outlining the inputs and outputs at various stages and the accompanying anthropological factors

attributes (Lechtman 1997; Lemonnier 1993). The *chaîne opératoire* framework focuses attention on the sequence of steps in artefact production and the embedded cultural choices. As such, it takes technological studies out of a black box that up to that point had diverted attention from the cultural dimensions of technological production in Western technological studies (Dobres 2000). More recently concepts such as 'materiality' have elaborated different elements of the *chaîne opératoire* approach (Jones 2004). As an analytical concept, materiality simultaneously considers the technological and anthropological factors of artefacts and technologies that produced them and falls within the broader social construction of technology paradigm (Killick 2004; Martinón-Torres and Rehren 2009b).

Another useful analytical concept is that of technological style or technology of practice which is specific to groups of people across space and time (Stahl 2009). Technological style and technology of practice situate the repertoire of metalworking within communities, making it possible to explore diachronic and synchronic variation. All this can be enveloped within the broader social constructivism movement in anthropology which seeks to consider both scientific and related cultural factors (Killick 2004). While global archaeology has now realized that dissecting technology into the scientific and magical is wholly unnecessary, studies of African metallurgy have long realized this. Indigenous African metallurgy is imbued with metaphors of birth and reproduction (Childs 1991; Ndoro 1991; Schmidt 2006); it is also a technical process involving thermochemically gaining elemental metal from ore (Childs and Herbert 2005). All these processes are rooted in society and social structure and entail both human–human relations as well as human material relations.

This book continues with this proud Africanist tradition of considering the quartet of mining, smelting, smithing and distribution simultaneously as a scientific and cultural process, but pushes a bit further. A central strand that runs through the chapters is the need to open conceptual boundaries and close gaps between the technological and sociological dimensions of metallurgy (cf. Schmidt 2009; also David 2012). Radiating from this centre is a strong focus on regional variation, the social engineering role of metals in society and the general societal implications emanating from the adoption of metallurgy. Examples from other areas of the world will, where possible, be brought into the discussion to forge a rich comparative perspective.

Organization of Work

This book considers metals and alloys worked in North African and sub-Saharan antiquity and explores variability in their working and uses from area to area within the wider archaeology of the continent. It seeks to communicate to both a global and local audience the key attributes of African metallurgy, including technological variation across space and time, methods of mining and extractive metallurgy and the fabrication of metals within and between Africa and the outside world. Although metallurgy was introduced at different points in time across the continent, the metals and alloys shaped in a significant and permanent way the development and organisation of metalworking societies and their destinies.

Chapter 2 discusses the origins and development of metal production and use in various parts of the African continent. Chapter 3 focuses on the anthropology and technology of mining ores worked in precolonial Africa. It shows that the techniques of extracting ore from the ground were a cultural venture, mediated by ancestors who made it possible to cross the boundary into the earth, to haul out its riches. Chapter 4 deals with the process of transforming through heat ore into metal during reduction, combining scientific with ethnographic observations. Smelting was socioculturally embedded and was an important transformative process that turned an object from nature (ore) into a cultural product (metal). The technology used across the continent varied according to local specifics such as geology, scale of production and many others. Chapter 5 discusses the fabrication of metals into objects and how metals acquired social and cultural roles. Chapter 6 discusses the broader issues associated with metals in society, including how they were distributed and their consequences for producers, consumers and society at large. It shows that overland links developed between communities on the continent, while sea links enchained communities to Indian Ocean and later Atlantic trading systems. This promoted urbanization at places such as the Swahili city states of East Africa, Great Zimbabwe in Southern Africa and, among others, ancient Ghana in West Africa. Chapter 7 emphasizes lessons for global anthropology that emanate from Africa.

References

Appadurai, A. (Ed). (1986). *The social life of things: Commodities in cultural perspective*. Cambridge: Cambridge University Press.

Bachmann, H. (1982). *The Identification of slags from archaeological sites*. Occasional Publication no 6. London: Institute of Archaeology.

Bayley, J., & Rehren, T. (2007). Towards a functional and typological classification of crucibles. In S. La Niece, D. Hook, & P. Craddock (Eds.), *Metals and mines: Studies in archaeometallurgy* (pp. 46–55). London: Archetype.

Bent, J. T. (1896). *The ruined cities of Mashonaland: Being a record of excavation and exploration in 1891*. London: Longmans, Green, and Co.

Beach, D. N. (1980). *The Shona and Zimbabwe, 900–1850: an outline of Shona history*. London: Heinemann.

Chikwendu, V. E., Craddock, P. T., Farquhar, R. M., Shaw, T., & Umeji, A. C. (1989). Nigerian sources of copper, lead and tin for the Igbo-Ukwu bronzes. *Archaeometry, 31*(1), 27–36.

Childs, S. T. (1991). Style, technology and iron smelting furnaces in Bantu-speaking Africa. *Journal of Anthropological Archaeology, 10*(4), 332–359.

Childs, S. T., & Herbert, E. W. (2005). Metallurgy and its consequences. In A. B. Stahl (Ed.), *African archaeology: Acritical introduction* (pp. 276–300). Oxford: Blackwell.

Childs, S. T., & Killick, D. (1993). Indigenous African metallurgy: Nature and culture. *Annual Review of Anthropology, 22,* 317–337.

Chirikure, S. (2007). Metals in society: Iron production and its position in Iron Age communities of Southern Africa. *Journal of Social Archaeology, 7*(1), 72–100.

Chirikure S. (2014). Geochemistry of ancient metallurgy: Examples from Africa and elsewhere. In H. D. Holland & K. K. Turekian (Eds.), *Treatise on Geochemistry*, Second Edition, vol. 14 (pp. 169–189). Oxford: Elsevier.

Chirikure, S., & Rehren, T. (2006). Iron smelting in precolonial Zimbabwe: Evidence for diachronic change from Swart Village and Baranda, Northern Zimbabwe. *Journal of African Archaeology, 4,* 37–54.

Chirikure, S., Heimann, R. B., & Killick, D. (2010). The technology of tin smelting in the Rooiberg Valley, Limpopo Province, South Africa, ca. 1650–1850 CE. *Journal of Archaeological Science, 37*(7), 1656–1669.

Chirikure, S., Manyanga, M., & Pollard, A. M. (2012). When science alone is not enough: Radiocarbon timescales, history, ethnography and elite settlements in Southern Africa. *Journal of Social Archaeology, 12*(3), 356–379.

Cline, W. B. (1937). *Mining and metallurgy in Negro Africa* (No. 5). Menasha: George Banta Publishing Company.

Cohen, C., van Buren, M., Mills, B., & Rehren, T. (2010). Current silver smelting in the Bolivian Andes: A review of the technology employed. *Historical Metallurgy, 44,* 153–162.

Craddock, P. T. (1995). *Early metal mining and production*. Washington, DC: Smithsonian University Press.

David, N. (2012). *Metals in Mandara mountains society and culture*. Trenton NJ: Africa World Press.

David, N., & Kramer, C. (2001). *Ethnoarchaeology in action*. Cambridge: Cambridge University Press.

David, N., Heimann, R., Killick, D., & Wayman, M. (1989). Between bloomery and blast furnace: Mafa iron-smelting technology in North Cameroon. *African Archaeological Review, 7*(1), 183–208.

de Barros, P. (1988). Societal repercussions of the rise of large-scale traditional iron production: A West African example. *African Archaeological Review, 6*(1), 91–113.

de Barros, P. (2013). A comparison of early and later Iron Age societies in the Bassar region of Togo. In J. Humpris & Th. Rehren (Eds.), *The World of Iron* (pp. 10–21). London: Archetype.

de Garis Davies, N. (1943). *The Tomb of Rekh-mi-Reâ at Thebes* (Vol. 2). New York: Arno Press.

DeCorse, C. R. (1992). Culture contact, continuity, and change on the Gold Coast, AD 1400–1900. *African Archaeological Review, 10*(1), 163–196.
Deer, W. A., Howie, R. A., & Zussman, J. (1992). *Rock-forming minerals: Vol 2 chain Silicates*. Longman.
Dewey, W. J. (1991). *Weapons for the ancestors*. Des Moines, University of Iowa, Department of Art History.
Dobres, M. A. (2000). *Technology and social agency: Outlining a practice framework for archaeology*. Oxford: Blackwell.
Edwards, D. N. (2004). *The Nubian past: An archaeology of the Sudan*. London: Routledge.
Emery, W. B. (1963). Egypt Exploration Society preliminary report on the excavations at Buhen, 1962. *Kush, 11,* 116–120.
Fenn, T. R., Killick, D. J., Chesley, J., Magnavita, S., & Ruiz, J. (2009). Contacts between West Africa and Roman North Africa: Archaeometallurgical results from Kissi, Northeastern Burkina Faso. In S. Magnavita, L. Koté, P. Breunig, & O. A. Idé (Eds.), *Crossroads/Carrefour Sahel: Cultural and technological developments in first millennium BC/AD West Africa* (pp. 119–146). Frankfurt am Main: Africa Magna Verlag. (Journal of African Archaeology Monograph Series 2).
Gale, N. H., & Stos-Gale, Z. A. (1982). Bronze Age copper sources in the Mediterranean: A new approach. *Science, 216*(4541), 11–19.
Garrard, T. F. (2011). (1989). *African gold: Jewellery and ornaments from Ghana, Côte d' 'Ivoire, Mali and Senegal in the collection of the Barbier-Mueller Museum*. Munich: Prestel.
Haaland, R. (2004a). Technology, transformation and symbolism: Ethnographic perspectives on European iron working. *Norwegian Archaeological Review, 37*(1), 1–19.
Hansen, B. (1986). The complementarity of science and magic before the scientific revolution. *American Scientist, 74*(2), 128–136.
Hauptmann, A. (2007). *The archaeometallurgy ofcCopper: Evidence from Faynan, Jordan*. New York: Springer.
Herbert, E. W. (1993). *Iron, gender, and power: Rituals of transformation in African societies*. Bloomington: Indiana.
Hoover, H. C., & Hoover, L. R. (trans.) 1950. *Georgius Agricola's De Re Metallica*. New York: Dover Publications.
Horton, M. C., & Middleton, J. (2000). *The Swahili: The social landscape of a mercantile society*. Oxford: Blackwell.
Huysecom, E., & Agustoni, B. (1997). *Inagina: l'ultime maison du fer/the last house of iron*. Geneva: Telev. Suisse Romande. Videocassette, 54 min.
Insoll, T. (2008). Negotiating the archaeology of destiny. An exploration of interpretive possibilities through Tallensi Shrines. *Journal of Social Archaeology, 8*(3), 380–403.
Jones, A. (2004). Archaeometry and materiality: Materials-based analysis in theory and practice. *Archaeometry, 46*(3), 327–338.
Joosten, I., Jansen, J. B. H., & Kars, H. (1998). Geochemistry and the past: Estimation of the output of a Germanic iron production site in the Netherlands. *Journal of Geochemical Exploration, 62*(1), 129–137.
Keen, J. (2013). Smelting: A sacred process. Observations of iron smelting in Madhya Pradesh, India. In J. Humpris & T. Rehren (Eds.), *The world of Iiron* (pp. 97–103). London: Archetype.
Killick, D. (2004a). Review essay: What do we know about African iron working? *Journal of African Archaeology, 2*(1), 97–112.
Killick, D. (2014a). Cairo to Cape: The spread of Metallurgy through Eastern and Southern Africa. In B. W. Roberts & C. Thornton (Eds.), *Archaeometallurgy in global perspective* (pp. 507–527). New York: Springer.
Killick, D. (2014b). From ores to metals. In B. W. Roberts & C. Thornton (Eds.), *Archaeometallurgy in global perspective* (pp. 11–45). New York: Springer.
Killick, D. J. (1990). *Technology in its social setting: Bloomery iron-smelting at Kasunga, Malawi, 1860–1940*. University Microfilms.
Killick, D., & Fenn, T. (2012). Archaeometallurgy: The study of preindustrial mining and metallurgy. *Annual Review of Anthropology, 41,* 559–575.

References

Kiriama, H. O. (1987). Archaeo-metallurgy of iron smelting slags from a Mwitu Tradition site in Kenya. *The South African Archaeological Bulletin*, 125–130.

Klemm, R., & Klemm, D. D. (2012). *Gold and gold mining in ancient Egypt and Nubia: Geoarchaeology of the ancient gold mining sites in the Egyptian and Sudanese eastern deserts*. New York: Springer.

Kusimba, C. M. (1993). *The archaeology and ethnography of iron metallurgy on the Kenya coast* Unpublished Doctoral dissertation, Bryn Mawr College.

Lane, P. J. (2005). Barbarous tribes and unrewarding gyrations? The changing role of ethnographic imagination in African archaeology. In A. B. Stahl (Ed.), *African archaeology: A critical Introduction* (pp. 24–54). Oxford: Blackwell.

Lechtman, H. (1977). Style in technology—some early thoughts. In H. Lechtman & R. Merrill (Eds.), *Material culture: Styles, organization and dynamics of technology* (pp. 3–20). New York: West.

Lemonnier, P. (1993). Introduction. In P. Lemonnier (Ed.), *Technical choices* (pp. 1–35). London: Methuen.

Levtzion, N. (1973). *Ancient Ghana and Mali* (Vol. 7). London: Methuen.

Levtzion, N., & J.F.P. Hopkins, J. F. P. (Eds.). (2000). *Corpus of early Arabic sources for West African history*. Princeton: Markus Wiener Publishers.

Martinón-Torres, M., Rehren, T., & Freestone, I. C. (2006). Mullite and the mystery of Hessian wares. *Nature, 444*(7118), 437–438.

Martinón-Torres, M., & Rehren, T. (Eds.). (2009a). *Archaeology, history and science: Integrating approaches to ancient materials*. Walnut Creek: Left Coast Press.

Martinón-Torres, M., & Rehren, T. (2009b). Post-medieval crucible production and distribution: A study of materials and materialities. *Archaeometry, 51*(1), 49–74.

Mauss, M. (1954). *The gift: Forms and functions of exchange in archaic societies* (No. 378). New York: WW Norton & Company.

Miller, D. (2002). Smelter and smith: Iron Age metal fabrication technology in Southern Africa. *Journal of Archaeological Science, 29*(10), 1083–1131.

Morton, G. R., & Wingrove, J. (1969). Constitution of bloomery slags: Part I: Roman. *Journal of the Iron and Steel Institute, 207*, 1556–1564.

Morton, G. R., & Wingrove, J. (1972). Constitution of bloomery slags: Part II: Medieval. *Journal of the Iron and Steel Institute, 210*, 478–487.

Mudenge, S. I. (1974). The role of foreign trade in the Rozvi empire: A reappraisal. *Journal of African History, 15*(3), 373–391.

Mudenge, S. I. (1988). *A political history of Munhumutapa c 1400–1902*. Harare: Zimbabwe Publishing House.

Muralha, V. S., Rehren, T., & Clark, R. J. (2011). Characterization of an iron smelting slag from Zimbabwe by Raman microscopy and electron beam analysis. *Journal of Raman Spectroscopy, 42*(12), 2077–2084.

Ndoro, W. (1991). Why decorate her. *Zimbabwea, 3*, 5–13.

Okafor, E. (1993). New evidence on early iron-smelting in Southeastern Nigeria. In T. Shaw, P. Sinclair, B. Andah, & A. Okpoko (Eds*.), The archaeology of Africa: Food, metals, and towns* (pp. 432–448). London: Routledge.

Pernicka, E., Begemann, F., Schmitt-Strecker, S., Todorova, H., & Kuleff, I., (1997). Prehistoric copper in Bulgaria. Its composition and provenance. *Eurasia Antiqua, 3*, 41–180.

Phillips, W. R., & Griffen, D. (2004). *Optical mineralogy: Non-opaque minerals*. New York: W. H. Freeman and Company.

Phimister, I. R. (1974). Alluvial gold mining and trade in nineteenth-century South Central Africa. *The Journal of African History, 15*(3), 445–456.

Pieterse, J. N. (1998). *White on black: images of white and blacks in western popular culture*. New Haven: Yale University Press.

Pleiner, R. (2000). *Iron in archaeology: the European bloomery smelters*. Archeologický ústav AVČR.

Pollard, A. M., & Bray, P. (2007). A bicycle made for two? The integration of scientific techniques into archaeological interpretation. *Annual Review of Anthropology, 36,* 245–259.

Pollard, A. M., Batt, C. M., Stern, B., & Young, S. M. M. (2011). *Analytical chemistry in archaeology*. Cambridge: Cambridge University Press.

Pringle, H. (2009). Archaeology. Seeking Africa's first iron men. *Science, 323*(5911), 200–202.

Radivojević, M., Rehren, T., Pernicka, E., Šljivar, D., Brauns, M., & Borić, D. (2010). On the origins of extractive metallurgy: New evidence from Europe. *Journal of Archaeological Science, 37*(11), 2775–2787.

Rehren, T., & Pernicka, E. (2008). Coins, artefacts and isotopes–archaeometallurgy and archaeometry. *Archaeometry, 50*(2), 232–248.

Rehren, T., Charlton, M., Chirikure, S., Humphris, J., Ige, A., & Veldhuijzen, H. A. (2007). Decisions set in slag: The human factor in African iron smelting. In S. La Niece, D. Hook, & P. Craddock (Eds.), *Metals and mines: Studies in archaeometallurgy* (pp. 211–218). London: Archetype.

Rickard, T. A. (1939). The primitive smelting of iron. *American Journal of Archaeology, 43*(1), 85–101.

Rothenberg, B. (1970). An archaeological survey of South Sinai: First Season 1967/1968, Preliminary Report. *Palestine Exploration Quarterly, 102*(1), 4–29.

Rothenberg, B. (1999). Archaeo-metallurgical researches in the Southern Arabah 1959–1990. Part 2: Egyptian new kingdom (Ramesside) to early Islam. *Palestine Exploration Quarterly, 131,* 149–175.

Scheel, B. (1989). *Egyptian metalworking and tools*. Oxford: Shire Publications.

Schmidt, P. R. (1978). *Historical Archaeology: A structural approach in an African culture*. Westport: Greenwood Press.

Schmidt, P. R. (1996). *The culture and technology of African iron production*. Gainesville: University Press of Florida.

Schmidt, P. R. (1997). *Iron technology in East Africa: Symbolism, science, and archaeology*. Bloomington: Indiana University Press.

Schmidt, P. R. (2009). Tropes, materiality, and ritual embodiment of African iron smelting furnaces as human figures. *Journal of Archaeological Method and Theory, 16*(3), 262–282.

Schultze, C., Stanish, C., Scott, D., Rehren, T., Kuehner, S., & Feathers, J. (2009). Direct evidence of 1,900 years of indigenous silver production in the Lake Titicaca Basin of Southern Peru. *Proceedings of the National Academy of Sciences, 106,* 17280–17283.

Srinivasan, S. (1994). Wootz crucible steel: A newly discovered production site in South India. *Institute of Archaeology, University College London, 5,* 49–61.

Stahl, A. B. (1994). Change and continuity in the Banda area, Ghana: The direct historical approach. *Journal of Field Archaeology, 21*(2), 181–203.

Stahl, A. B. (2001). *Making history in Banda: Anthropological visions of Africa's past*. Cambridge: Cambridge University Press.

Stahl, A. B. (2009). The archaeology of African history. *The International Journal of African Historical Studies, 42*(2), 241–255.

Summers, R. (1969). *Ancient mining in Rhodesia and adjacent areas*. Salisbury: Trustees of the National Museum of Rhodesia.

Tripathi, V. (2013). An ethno-archaeological survey of iron working in India. In J. Humpris & T. Rehren (Eds.), *The world of Iiron* (pp. 1094–1115). London: Archetype.

Van der Merwe, N. J., & Avery, D. H. (1987). Science and magic in African technology: Traditional iron smelting in Malawi. *Africa, 57*(2), 143–172.

Vansina, J. M. (1985). *Oral tradition as history*. Madison: University of Wisconsin Press, New York: Springer.

Zangato, E., & Holl, A. F. (2010). On the iron front: New evidence from North-Central Africa. *Journal of African Archaeology, 8*(1), 7–23.

Chapter 2
Origins and Development of Africa's Preindustrial Mining and Metallurgy

Introduction

From antiquity, Africa has been simultaneously a continent of similarities and differences. Geographically and to some extent culturally, Africa can be divided into discrete regions: West, Central, East, and Southern Africa, as well as the Horn and North Africa including Egypt. Egypt, North Africa and the Horn have a long history of participation in the metallurgical, ceramic, glass and other high temperature technological traditions of the Near East and the Mediterranean worlds. Other regions, primarily in sub-Saharan Africa, form a distinctly different area, which, although interacting with North Sahara, particularly after 500 BC (see Stahl 2014a and references therein), forms a distinct cultural and technological block.

Egypt and adjacent regions closely mimic the metallurgical trajectories of the nearby Middle East. Egyptian metallurgy started with the working of copper around 4000 BC. By 3000 BC, the Bronze Age was fully established with iron appearing much later in the last millennium BC (Scheel 1989). Because Egypt had cultural interactions with regions to the south of the Nile, metallurgy was established in Nubia by 2600 BC (Emery 1963). Iron smelting appeared much later in Egypt (c. 600 BC; Scheel 1989) when compared to the rest of the Middle East and was established even later (c. 500 BC) at places such as Meroe in the Sudan (Rehren 2001). In North Africa, the Phoenician settlements at Carthage were established by 600 BC (Fig. 2.1). The development of metallurgy in Carthage is not clearly understood, but it is clear that by 600 BC or shortly after, Carthaginians worked iron, copper and bronze (Alpern 2005).

Sub-Saharan Africa differs from this picture in that metallurgy in this part of the continent began with the working of iron and in some cases iron and copper (Holl 2009). This is especially true in West Africa, Central Africa, East Africa and Southern Africa. The advent of metallurgy in sub-Saharan Africa is a highly contentious topic because for every possibility, there are two or more contradictions (Craddock 2010). Metallurgy in West, East and Central Africa began sometime between 800 and 400 BC in the radiocarbon black hole created by fluctuations in atmospheric concentration of radiocarbon (Clist 2013; Killick 2004a). In Southern Africa,

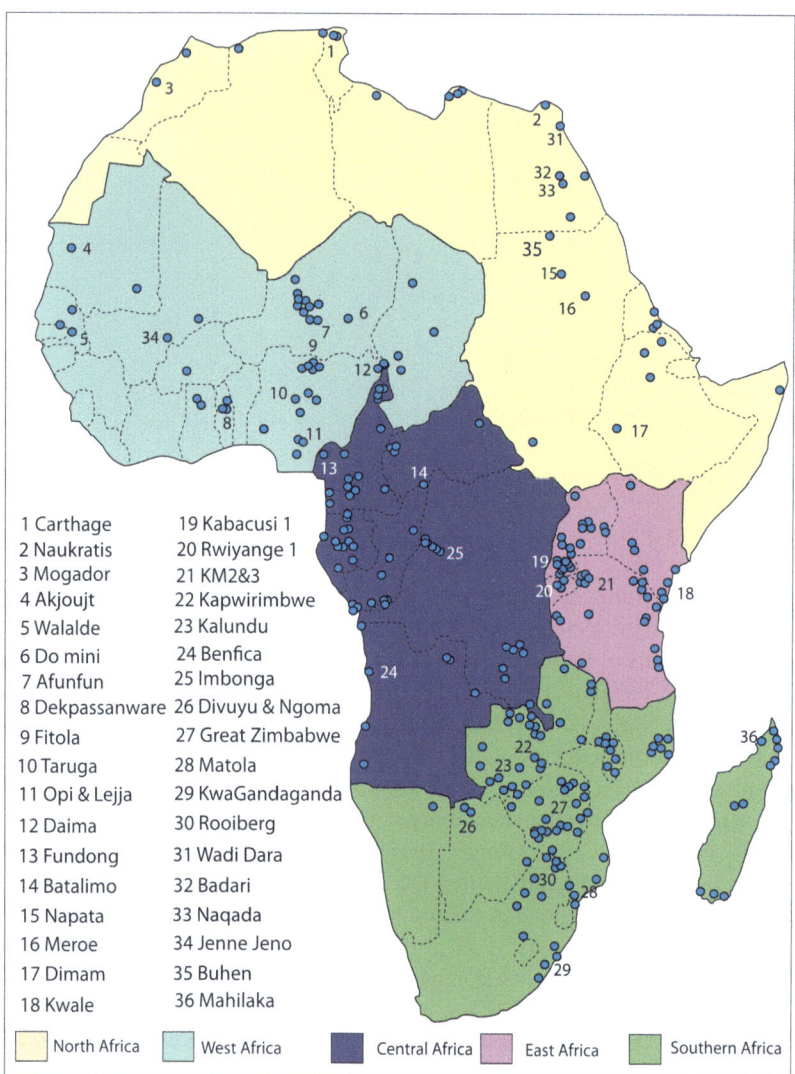

Fig. 2.1 Map of Africa showing metalworking sites with some of the most important highlighted by number

metallurgy only appeared with the advent of agriculturalists early in the first millennium AD (Phillipson 2005). After almost a thousand years, bronze and gold made their appearance in sub-Saharan Africa when the region was directly integrated into the Islamic trade via the Sahara and the Indian Ocean littoral. This difference with the picture north of the Sahara precipitated the development of a raging and largely unresolved debate regarding the origins of sub-Saharan African metallurgy, particularly whether it is local or external in origin (Alpern 2005). Whatever the case maybe, the differences in the adoption of metallurgy in Africa's different regions provide important lessons for innovation, technology transfer and cultural interaction. Once established, metallurgy was neither static nor homogenous throughout antiquity. It developed locally and regionally, creating a very richly varied history of local innovation and cross-cultural borrowing.

Origins of Metallurgy in Egypt and Adjacent Areas

The earliest evidence for metallurgy in Africa comes from the Nile Delta in Egypt and is associated with the Maadi culture dating between 4000 and 3200 BC (Killick 2014a; Scheel 1989). Evidence suggests that copper substituted for flint as the raw material for making heavy-duty tools during this period. The paucity of copper deposits in this area coupled to its proximity to the Sinai Desert and Southern Jordan presents a persuasive but untested hypothesis that the copper of the Arabah Desert was used by Maadi people. Elsewhere in Egypt, rare ornaments and implements of copper metal were recovered in the middle Nile during the Badarian period (ca. 4400–4000 cal BC). However, no archaeometallurgical studies were carried out to determine whether they were made of smelted or native copper (Killick 2014a). Still in the middle Nile, although copper oxides were used in the Naqada I period (4000–3500 cal BC), heavy-duty copper tools such as axes and blades were more common during the Naqada II phase (3500–3200 cal BC) (Scheel 1989). The ore used to make these objects probably came from the lower Nile near Nubia. Gold and silver also appear at low frequency in Naqada II graves (Midant-Renes 2000). It is possible that some if not all of this gold came from the Eastern Desert and later from Nubia (Klemm et al. 2003).

Copper and gold artefacts initially appeared in lower Nubia (the region between the First and Second Cataracts) in graves of the Middle A Group, which are dated from ca. 3600–3300 cal BC (Killick 2014a). These are associated with Naqada pottery and other items of Egyptian provenance, suggesting that they too were imported. By 3000 cal BC, copper beads, awls and pins were found as far south as the Third Cataract. Interestingly, it appears as if all cutting implements were still made of stone. The earliest evidence of the production of metals in Nubia is from Old Kingdom contexts (ca. 2600 BCE) at Buhen (Emery 1963) and within the temple precinct further upstream at Kerma, in contexts dated by radiocarbon to 2200–2000 cal BC. Ancient Egyptians forged meteoric iron (iron in its native state) from c. 3000 BC onwards to produce beads and other decorative items (Rehren et al. 2013). Indeed, sporadic iron objects were found at Egyptian sites, but it is generally accepted that iron smelting began in Egypt after its invasion by the Assyrians in 691 BC. Iron smelting then gradually filtered down the Nile to Kerma, Meroe and other places and was well established by c. 500 BC. Craddock (2010) argues that given the antiquity of its metallurgy, Kerma is a possible source of sub-Saharan metallurgy but more research is required to substantiate this thinking.

The Phoenicians are credited with introducing knowledge of metallurgy to North Africa, particularly to modern-day Tunisia and Libya. Around 1101 BC, the Phoenicians established the trading port of Utica in Tunisia and by 814 BC had established Carthage nearby (Alpern 2005). There is a great deal of debate regarding the metallurgical history of Carthage, but it is clear that iron was worked together with copper and bronze by 300 BC. Alpern (2005) cites unsubstantiated reports of iron smelting at Carthage dating to 800 BC. Unless corroborated by written texts, this dating too may be affected by the radiocarbon 'black hole' where the calibration curve flattens between cal 800 and 400 BC resulting in uncertain dates (Killick 2014a) and, like similar dates elsewhere in Africa, must be treated with caution. Seemingly, Phoenician ventures into the western Mediterranean were motivated by a desire to identify sources of gold, silver, copper and tin for trade purposes. This was crucial because the Egyptians had virtual monopoly over the gold from Nubia and the Eastern desert. Although copper is available at Akjoujt in Mauretania and tin in Niger's Aïr Mountains, it seems that Phoenicians never knew of these sources, preferring the tin of Cornwall that is believed to have featured in Carthaginian trade (Alpern 2005). Carthage features strongly in debates over origins of sub-Saharan metallurgy, with proponents of external origins speculating that it may have been a conduit in knowledge transfer. I return to this point after presenting the evidence for early metallurgy in sub-Saharan Africa.

Ethiopia and Eritrea are poorly understood as far as the development of metallurgy is concerned (Mapunda 1997; Phillipson 2005). It has been suggested that the Horn of Africa follows the progression witnessed in Egypt, Nubia and Arabia. As such, gold, copper, and silver and bronze were known in Ethiopia by the last centuries BC. Aksum witnessed the height of its power from the early first millennium AD and minted its own coinage in gold and silver (Phillipson 2005, p. 230). The beginning of iron working in Ethiopia was also late relative to adjacent regions, starting around cal 300 BC (Mapunda 1997). Indeed, the available evidence suggests close interaction between the Kingdom of Kush in its various stages and the Horn of Africa on the one hand and Egypt and the Mediterranean world on the other via the Red Sea trade.

Metal from Somewhere: On the Origins of Metallurgy in West, Central and East Africa

In the studies of Africa's later prehistory, no topic evokes as much debate and emotion as the origins of sub-Saharan metallurgy (Alpern 2005; Zangato and Holl 2010 and responses therein). When compared to the Middle East and the adjacent Balkans, which are widely believed to be independent centres of metallurgy (Radivojević et al. 2010), sub-Saharan metallurgy started simultaneously with the working of copper and iron (Van der Merwe and Avery 1982) (Table 2.1 & Fig. 2.1). The pathway to metallurgy in Middle Eastern and Balkan centres of metallurgical origins, as well as in Egypt, began with the intentional heating of oxide and carbonate copper

Table 2.1 shows some of the earliest dates for the appearance of metallurgy in Africa. Calibrated using OxCal version 4.2.3 Bronk-Ramsey (2013) and IntCal13 (Reimer et al. 2013)

Site name	Lab nos.	Uncalibrated dates	Calibrated dates at 95 % confidence interval	Sources
Termit Massif, Niger				
Do Dimmi 16 a M Do Dimmi 15 a F	UPS IFAN	2590±120 2630±120	978–404 BC 1031–410 BC	Person and Quenchon 2004, p. 122
Gara Tchia Bo 48 E	Pa 810	3260±100	1770–1290 BC	Person and Quenchon 2004, p. 122
Gara Tchia B 48 W	Pa 811	3265±100	1775–1294 BC	Person and Quenchon 2004, p. 122
Tchire Ouma 147	Pa 320	3300±120	1895–1370 BC	Person and Quenchon 2004, p. 122
Termit Ouest 96 b M	Pa 481	3100±100	1611–1107 BC	Person and Quenchon 2004, p. 122
Termit Ouest 8-b	Pa 688	2880±120	1322–819 BC	Person and Quenchon 2004, p. 122
Nsukka Region, Nigeria				
Opi	OxA-3201	2305±90	596–166 BC	Okafor 1993, p. 347
	OxA2691	2170±80	396–40 BC	Okafor 1993, p. 347
	Oxa3200	2080±90	361 BC–70 AD	Okafor 1993, p. 347
Lejja	Ua 34416	1715±35	244–398 AD	Eze-Uzomaka 2013
	Ua 34417	2370±40	545–380 BC	Eze-Uzomaka 2013
	Ua 34415	4005±40	2631–2458 BC	Eze-Uzomaka 2013
Taruga	BM938	2541±104	846–403 BC	Calvacoressi and David 1979
Taruga	BM942	2291±123	596–98 BC	
Togo				
Dekpassanware	Beta 252674	2970±40	1297–1051 BC	De Barros 2013
Cameroon				
Olinga	Beta 31414	2820±70	1131–827 BC	Essomba 2004, p. 140

Table 2.1 (continued)

Site name	Lab nos.	Uncalibrated dates	Calibrated dates at 95% confidence interval	Sources
	Ly4978	2380±110	792–347	Essomba 2004, p. 140
	Ly4979	1954±250	544 BC–590 AD	Essomba 2004, p. 140
	Beta 31412	1860±70	2–345 AD	Essomba 2004, p. 140
Central African Republic				
Obui	Pa 2223	3645±35	2136–1921 BC	Zangato and Holl 2010
Obui	Pa 2130	3635±35	2058–1903 BC	Zangato and Holl 2010
Gbabiri	Pa 1446	2670±40	898–797 BC	Zangato and Holl 2010
Rwanda				
Rwiyange	HV 1296	2250±125	593–20 BC	Van Grunderbeek et al. 2001
Mozambique				
Matola	R1327	1880±50	19–246 AD	Huffman 2007, p. 163
	St8546	1720±110	70–550 AD	Huffman 2007, p. 163
South Africa				
Silver leaves	Pta 2360	1760±50	137–386 AD	Huffman 2007, p. 163
	Pta 2459	1700±40	246–416 AD	Huffman 2007, p. 163
Broederstroom	KN 2643	1600±50	344–569 AD	Huffman 2007, p. 163
		1350±80	547–880 AD	Huffman 2007, p. 163
Zimbabwe				
Mabveni	SR79	1380±110	425–886 AD	Huffman 2007, p. 163
Gokomere	SR26	1420±120	386–886 AD	Huffman 2007, p. 163

ores in temperature- and environment-regulated apparatuses to gain a usable product (Craddock 1995; Pernicka et al. 1997; Radivojević et al. 2010; Scheel 1989). The Bronze Age started with the working of arsenical copper followed by the alloying of tin with copper to produce bronze, with the more complicated iron appearing around 1500 BC (Tylecote 1976; Craddock 2000). Egypt and areas under its influence along the Nile broadly followed this trajectory of copper, bronze and iron transition. Gold and other metals such as lead were also worked during this time, such that a long-distance trade had evolved by 2000 BC. Despite its advantages, iron was not universally accepted in the Middle East because Egypt only fully embraced it around 700 BC, more than six centuries after its adversaries, neighbours and trading partners adopted it (Craddock 2000; Holl 2000). The Cushite Egyptian Pharaohs were defeated by iron-armed Assyrians in 691 BC. This supports the argument that the adoption of metallurgy, such as technology in general, is culturally mediated; no matter how many perceived advantages there are, society determines what is and what is not acceptable.

The path to metallurgy in the Middle East and adjacent regions indicates that discovery and innovation followed the easiest methods through which the very first metals could be worked (Craddock 2010). Such a picture partly intersects with the laws of physics and chemistry as summarized by the Ellingham diagram (Fig. 2.2) which presents the temperature at which oxide ores are reduced to metal in relation to the levels of carbon monoxide sufficient for reduction. According to Killick (2014b), pioneer metals such as copper and tin could be easily reduced at low temperatures, while latecomers such as iron required much higher temperatures and delicate control of furnace atmosphere to reduce their ores. Following a technical logic, this seems to account for why copper and tin were smelted earlier than iron. However, it is not just a temperature issue, but also one of redox–carbon monoxide is not strong enough to reduce 'modern' metals (Th. Rehren pers comm 2014).

The Ellingham diagram does not fully explain the sequence of metallurgical innovation in antiquity (Killick 2014b, p. 35). For example, metals such as cobalt and nickel (Fig. 2.2) have a lower melting point when compared to iron and are reducible at even lower temperatures. Yet, they were only smelted in the nineteenth century. In fact, nickel is more abundant than copper in the earth's crust, while cobalt is more abundant than lead (Killick 2014b). There are many possibilities that account for why nickel and cobalt were not smelted in the known centres of metallurgical origins. The most important one is that nickel and cobalt oxides are quite soluble in water and thus are almost never found in gossans (intensely oxidized, weathered and exposed/upper part of an ore deposit or mineral vein), making the fact that nickel and cobalt oxides are relatively easy to reduce irrelevant and there were no oxide or carbonate ores of these elements available (Killick 2014b). This also demonstrates that laws of physics and chemistry do not always fully explain the evolution of cultural phenomena. In fact, the laws themselves are cultural phenomena which were discovered at various points, explaining why most metals were discovered much later, and most of them not in any order that respects the known affinities between them.

In sub-Saharan Africa, tin, bonze and gold were worked more than a millennium after iron and copper were introduced. This period coincided with the integration of the subcontinent into the fledging long-distance trading network rooted in the Persian Gulf and the Indian subcontinent (Miller and Van der Merwe 1994). The big question, therefore, is where did knowledge of metalworking in sub-Saharan Africa

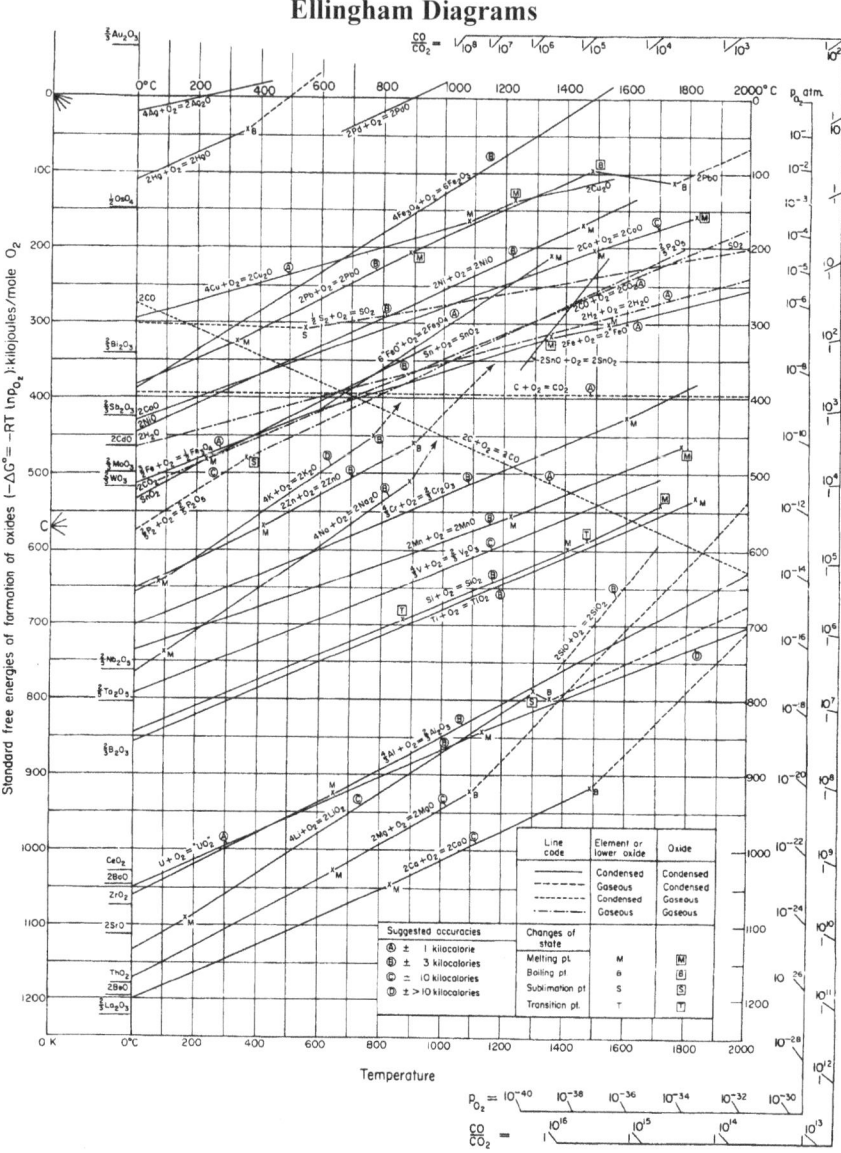

Fig. 2.2 Ellingham diagram shows the ease with which metals and sulphides can be reduced. The position of the line for a given reaction on the Ellingham diagram shows the stability of the oxide as a function of temperature. Reactions closer to the *top* of the diagram are the most 'noble' metals (for example, gold and platinum), and their oxides are unstable and easily reduced. Moving towards the *bottom* of the diagram, metals become progressively more reactive and their oxides harder to reduce

originate? Is it local or external in origin? As a follow on, what are the mechanisms for the dispersal of knowledge of metallurgy across the sub-Saharan latitudes? Based on thermodynamic theory and the lateness with which the continent embraced metallurgy in comparison with regions such as the Middle East (Fig. 2.2), one school of thought argues that knowledge of West and East African metalworking was external in origin (Killick 2004a; McIntosh and McIntosh 1988; Phillipson 2005).

The fulcrum of this position is that it is very difficult to start smelting a technically complicated metal such as iron without exposure to easier metals or comparable pyrotechnology as illustrated by the Middle Eastern trajectory (Craddock 2000; Killick 2014a). Furthermore, there are technical problems with early dates for West and East African metallurgy which may have been affected by the old wood problem (Alpern 2005) and the problems with calibration for dates falling between 800 and 400 BC. The old wood problem emanates from the fact that some of the trees in sub-Saharan Africa lived for long periods due to the gradual desertification of the Sahara between 4500 and 2000 BC (Childs and Herbert 2005; Killick 1987). If charcoal from old wood was used for smelting and subsequently dated by archaeologists, the dates produced reflect when the trees died and not necessarily the metalworking episodes. This has led to a rejection of most of the early dates for African metallurgy, some of which were additionally compromised by uncertainty of contexts (Clist 2013). Then there is the fact that the calibration curve flattens between 2300 and 2600BP which gives a very long tail between 800 and 400 BC and with that a great deal of uncertainty (Alpern 2005). It has long been advocated that researchers must use alternative dating techniques such as luminescence dating (Killick 2004a) though few have heeded the call (Darling 2013).

If West, Central and East African metallurgy emerged from outside, as posited by the external origins theory, what transmission routes did it follow? The site of Akjoujt in Mauritania has yielded copper working objects dating to 800 BC, suggesting a possible introduction from Morocco and a copper to iron transition (Miller and Van der Merwe 1994). However, earlier thoughts suggested that Egypt and Carthaginian settlements in North Africa were possible conduits for a north-to-south transmission of knowledge (Childs and Herbert 2005). Alpern (2005) believes that iron working was well established at Carthage by c. 800 BC, making it possible that West African metallurgy diffused from there. The only problem on the basis of current knowledge is that the dating is not universally agreed on and that iron became established in Egypt after the invasion by Assyrians in 691 BC (Scheel 1989). Furthermore, the evidence for the appearance of iron in Carthage postdates that of the supposedly receiving areas of West Africa (Darling 2013: 158; Eze-Uzomaka 2013, p. 4). Another conundrum is the many outward differences between the furnace types used in Egypt, Carthage and other possible source areas when compared to those utilized at places such as Taruga in Nigeria (Tylecote 1975). If the sub-Saharans obtained knowledge of metallurgy from Carthage, Egypt or somewhere, then the rapidity with which they adapted the technologies to the local situations, without evident experimentation, thereby distinguishing their technology from its sources at the same time they were adopting it, is remarkable.

These anomalies became fodder for viewpoints that consider African metallurgy to be local in origin. The local origins hypothesis contends that because Africa has always been a centre of technological development throughout human history, there is no reason why metallurgy could not have been independently developed here.

More importantly, various communities on the African continent were aware of the transformative potential of fire since the middle Pleistocene times. The local origins viewpoint seemed to gain momentum in the late 1970s and early 1980s when Danilo Grébénart and his team excavated significant sites with evidence of early metallurgy in the Agadez region of Niger (Grébénart 1987). The dating evidence, when combined with archaeometallurgical analyses of remnant furnaces and slags, seemed to indicate that there was an earlier Copper Age, named Copper 1 (2000–1000 BC), followed by a later Copper 2 (1000 BC) phase. According to Grébénart (1987), iron working started in the Copper 2 period, suggesting that in similar fashion to other areas in the Old World, African metallurgy started with an apprenticeship phase of copper working, followed by smelting of the more technically complicated iron. This seemed to refute the argument that Africans could not have developed metallurgy independently because they lacked experience with an easier metallurgy. However, a meticulous re-investigation of the Agadez material by Killick et al. (1988) demonstrated that what were thought to be remnant furnaces associated with Grébénart's Copper 1 period were vitrified tree stumps. The re-examination further highlighted that all reliable evidence of metallurgy dated to the Copper 2 period, later than 1000 BC. For a while, the critique of this evidence seemed to tilt the pendulum into the direction of external origins.

More recently, indications from the work carried out by Zangato and others in Central Africa seem to challenge again the external origins thinking. Excavations at places such as Ôboui in the Central African Republic revealed artefacts and forges which were dated to a much earlier time period, between 2300 and 1900 cal B.C.— long before the Anatolians were working iron (Zangato and Holl 2010). Archaeometallurgical studies of the microstructure of iron objects revealed that they were made of bloomery iron. This Central African evidence generated intense debate, with critics arguing that although the dates formed a nice cluster, they were probably from old wood. Furthermore, it was argued that given the acidic nature of soils in tropical Africa, the iron objects seem remarkably well preserved for their age (Clist 2013). Other authorities such as Craddock (2010) dismissed the possibility that Africa started its own metallurgy with iron, suggesting that this is about as likely as a baby walking without first crawling.

While scholars continue to debate the relevance of these dates, a set of dates from the Leija sites in Nsukka Nigeria has not yet been considered in full. These dates are Ua-34415, 4005±40, and Ua-37422, 3445±40 (Eze-Uzomaka 2013). The Leija date Ua-34415 calibrates to the second millennium BC and was obtained from charcoal embedded in slag in a stratified context over a meter deep. Clist (2013) notes that the Nsukka dates are some of the best in terms of association between the dated material and the events of metalworking. Also in Nigeria, Darling (2013) dated samples of fired slag pit furnaces at the Durham Thermoluminescence Laboratory, producing very early dates of 2400 BC±1100 for Fitola (Dur TL57–2AS) and 1400 BC±850 (Dur TL57–3AS) for the site of Matanfada in the Hausaland area of Northern Nigeria. The Durham TL dates require some comment because they have unusually very high error terms. According to Darling (2013), the laboratory could not find any sources of error and control samples are currently being dated. Until this dating is properly resolved, the Fitola dates must be viewed with caution.

A point that begs deeper consideration is that the Leija sites seem to be far away from Phoenician influence and considerably pre-date Carthage and other centres. This makes a strong statement that dates alone will probably never solve the origins issue. What will happen if luminescence dates from Leija and other sites also produce second millennium BC dates as in the case of Darling (2013)? Are we prepared as scholars to accept that there is nothing wrong with iron smelting starting independently in West Africa? Yes, iron smelting may be complicated, but there is so much diversity in furnace types—including some very rudimentary bowls simply made of banana stems (see Celis and Nzikobanyanka 1976)—that suggest that iron can be reduced anywhere provided temperature and air were sufficient. As such, perhaps the argument that iron is complex to smelt is a presentist assumption. In any case, the laws of physics and chemistry, which are so lucid today, were mostly hap-hazard before the nineteenth century such that today people tend to think that discovery and innovation in the past followed the nice and neat picture depicted in the periodic table of elements. Indeed, smelting started with less complicated metals but skipped nickel and cobalt and jumped to iron in Eurasia (Killick 2014b). Perhaps if we shed off our presentist arguments, it is possible that iron was independently developed in Africa. The contexts favoring the development of certain technologies are very different. That being said, further work is required to produce more dates, interrogate contexts of recovery and explore possible interconnections between the various parts of Africa to increase the confidence in the research community.

In East Africa, van Grunderbeek et al. (2001) and Schmidt (1997) have produced important evidence relating to the beginning of metallurgy in this region. Despite being close to the Sudan, which was heavily influenced by Egypt, it seems that the development of metallurgy in the Great Lakes Region emerged as something remarkably different (Humphris and Iles 2013; Killick 2014a). The Great Lakes region has traditionally been seen as one of the independent centres for the evolution of metallurgy in sub-Saharan Africa. Interestingly, van Grunderbeek et al. (2001) have backed away from the dates of >2800 BP from Rwanda and Burundi because one cannot prove that these dates are not contemporary with the furnaces (Killick 2014a). On his part Schmidt (1997, p. 14) has disavowed the three oldest dates (>3000 BP) from Buhaya. There remain, however, at least four radiocarbon dates between ca. 2350 BP and ca. 2650 BP, each in good association with furnaces and on charcoal of small diameter, which should eliminate the possibility of an 'old wood' error. However, these all fall within a well-known 'black hole' in the radiocarbon calibration curve, and thus, all calibrate at two sigma to a calendar range of approximately 800–400 cal BCE. This is exactly the same range of calibrated age as for the earliest radiocarbon dates for iron slag at Meroe (Shinnie 1985; Rehren 2001) and appears to be contemporary with some West African and Carthaginian dates (Alpern 2005; Holl 2009).

As things stand, there is no consensus regarding the origins of metalworking in Africa—the knowledge came from somewhere, but that somewhere could be within or outside the continent. This is because the two competing paradigms have potent gaps which cannot be overlooked, regardless of how persuasive their proponents are. What Africa has had are very good reviews with a lot of common sense that unfortunately will not decisively solve the problem at hand. As such, and as

Killick (1987, 2014a) has long advocated, Africa requires more research, backed up by careful and robust dating programs to understand the origins and dispersal of metallurgy. One such fruitful area is Nigeria where good association has been established between the dated material and iron smelting. So far, evidence from Carthage presented by Alpern (2005) suggests an early Phoenician presence and metallurgy around 800 BC. However, as traders, it is not clear why the Phoenicians could not take advantage of mineral wealth in West Africa and why they would introduce only iron and copper and not bronze when local sources of tin were available in Niger. The old wood problem, while making much sense, must be further investigated to establish the time frames by which it affects the usability and acceptability of early dates within the context of limitations and error limits of chronometric dating techniques. Clist (2013) argues that it is prudent to identify the tree species responsible for the charcoal being dated, for it may assist with interpretation. The only challenge is that charcoal is chemically inert and can thus persist for millennia beyond the life of trees that produced it.

While the origins debate largely concerns West, East and Central Africa, Southern Africa appears to be relatively controversy free. Here, it is acknowledged that metallurgy consisting of iron and copper working was introduced by the southward migration of Bantu-speaking people early in the first millennium AD (Kiriama 1987; Pwiti 1996). As with the pattern in West and Central Africa, gold, tin and bronze appeared after the integration of the region into the long-distance trade network.

Why are Iron and Copper Earlier than Gold and Bronze in Sub-Saharan Africa? Some Provocative Thoughts

Whatever the source of sub-Saharan Africa's preindustrial metallurgy, one telling observation is that gold, tin, and the alloys bronze and brass were only worked after 700 AD when the region was integrated into the trading system centred on the Indian Ocean (including the Persian Gulf) (Miller and Van der Merwe 1994). What prompts deeper thought is the observation that the so-called source areas north of the Sahara already knew how to work bronze, tin and gold, making the question why it took a millennium before Africans adopted these metals and alloys a relevant one. Egyptian art, dating to 2500 BC and much later, shows black Africans working and smelting copper. These individuals probably came from the southern neighbors of Egypt, and indeed, ebony and other commodities from black Africa also ended up in Egypt and the broader Middle Eastern region (Pieterse 1998). Therefore, connections existed between sub-Saharan Africa and the Middle East via North Africa (Wilson 2012) and down the Nile (Edwards 2004). It is difficult not to accept that the parts of sub-Saharan Africa interacting with Mediterranean Africa at least knew about metalworking. During Roman times, rock art in Mauretania when coupled with compositional analyses of copper objects indicate the existence of indirect interaction between parts of West and North Africa (Fenn et al. 2009). The only complication is that the earliest dates for metallurgy (c. 800–400 BC), on the basis of current evidence, seem to be from Niger, Nigeria and the Central African Republic, and not from any of the supposed conduit areas. Indeed, the early dates from Akjoujt, which also fall into the radiocarbon black hole, and objects stylistically related to those from Morocco may demonstrate contact and confirms that there was a flow of ideas between regions adjacent to North Africa. If that is the case, why did this not happen in Niger, or Great Lakes East Africa which is closer to areas that had undisputed evidence of metallurgy? Again, this underscores the point that Africa is poorly known as far as metallurgical innovation is concerned, placing most of the existing knowledge on the edge of speculation. Other questions that arise are if knowledge of metalworking originated from Middle Eastern communities which were traversing distant territories in search of precious metals as early as 2000 BC, why did they not take advantage of the vast mineral wealth in sub-Saharan Africa as their successors did a thousand years later? And, why did Africa not keep on borrowing technologies after the establishment of increased contact between the different areas? Several reasons may account for this but it is doubtful whether the late development of trade in precious metals was a question of delayed consumption on the part of Middle Eastern communities. It seems most likely that Middle Easterners either had rich alternative sources of precious metals or they were ignorant of the metallurgical resources of sub-Saharan Africa which strongly questions the depth and strength of the diffusionist argument for the origins of African metallurgy.

This paradox prompted Cline (1937, p. 11) to comment that 'it is curious that the metal most active in drawing the attention of Arab and European worlds to Negro Africa was the one which the Negroes themselves generally value the least'. Certainly, this situation has something to do with the dynamics of values and technology transfer. In much later periods, gold played an important role in the value system of the Asante in present-day Ghana while copper was more valuable in many Central African societies to the extent that it was the 'red gold of Africa' (Herbert 1984). It seems, in most cases, that technology transfer and innovation was a complex endeavor based on serendipity and neither followed 'rule books' nor was it premised on 'common sense' because some 'advanced' technologies appeared, only to disappear and be rediscovered thousands of years later. Thornton and Rehren (2009) discuss the engineering and geochemical characteristics of a fourth millennium BC highly refractory copper working crucible recovered at Tepe Hissar, North-Eastern Iran. Although demonstrating advanced refractoriness and containing material properties sought after thousands of years later, Tepe Hissar crucibles were neither widely adopted nor did their manufacture last for very long (Thornton and Rehren 2009). What is interesting is that such a technology only surfaced in Europe, in a completely different context, in the postmedieval period (Martinón-Torres et al. 2006; Martinón-Torres and Rehren 2009). In sub-Saharan Africa, the Igbo Ukwu bronzes—arguably one of the finest achievements in metal casting the world over and dating back to the late first millennium AD (Shaw 1970)—form another example of advanced metallurgical techniques that surfaced and disappeared, only to emerge again centuries later (Chikwendu et al. 1989). The Yoruba of Southeastern Nigeria independently re-invented the technology of glass making between the eleventh and thirteenth centuries AD (Lankton et al. 2006). These examples seem to indicate that the appearance of seemingly advanced technologies in various places is a random variable, and possibly one with varying combinations and permutations. Presumably, the reason why, within culture-specific contexts, technologies can appear and disappear, is that their acceptance passes through cultural filters that determine whether or not they are acceptable (Thomas 1991). Demography is also a factor—if more people take on an innovation, it becomes dominant. Otherwise, it disappears. However, continuities and discontinuities in knowledge transmission/apprenticeship could presumably be a factor as well.

Conclusion

Although metallurgy runs deep in the pulse of African empires and kingdoms from Dynastic Egypt to Kush, ancient Ghana, Mali, the Swahili towns of East Africa and Great Zimbabwe to mention a few examples (Fenn et al. 2009; Horton 1996; Levtzion 1973; Pikirayi 2001; Summers 1969), the adoption of metallurgy was very different in these areas. In Egypt, Nubia, Ethiopia, Eritrea and North Africa, the metals known include copper, gold, silver, iron, lead, mercury and tin. This contrasts with sub-Saharan Africa where only iron and copper were known for 1000 years before the adoption of tin, gold and possibly silver. Interestingly, in Southeast Asia, the picture neatly follows that in the Middle East, raising the question why is sub-Saharan Africa different if it also received its knowledge from the same source—the Middle East?

Perhaps it is now time to move away from the presentist argument that iron smelting is complex. It may be today, but may not have been to ancient sub-Saharans. The argument that smelting must progress from less to more complicated metals contradicts even the way chemistry developed, which neither followed rules of simplicity nor easiness given that most elements with known chemical relationships and affinities were discovered at different times and often by accident.

The main take-home message from this chapter is that there are many unknowns as far as the origins of African metallurgy are concerned. It is impossible on the basis of current evidence to argue definitively for or against local and external origins. More research is required to produce new insights, backed up by well resolved dating and good association between the dated events and metal production. Fragmentary as our knowledge of it is, the origins of African metallurgy is of global importance because it highlights that the adoption of the technology was neither homogenous nor straightforward. Instead, it followed different contours mediated by cultural, economic, political and other factors. This book, however, is about the nature of the technology, its evolution over time and the associated sociocultural permutations and impacts within a global picture.

References

Alpern, S. B. (2005). Did they or didn't they invent it? Iron in sub-Saharan Africa. *History in Africa, 32*, 41–94.
Bronk Ramsey, C. (2013). OxCal Program, version 4.2 (http://c14.arch.ox.ac.uk/oxcal.html), Radiocarbon Accelerator Unit. University of Oxford.
Calvocoressi, D., & David, N. (1979). A new survey of radiocarbon and thermoluminescence dates for West Africa. *Journal of African History, 20*(1), 1–29.
Celis, G., & Nzikobanyanka, E. (1976). *La métallurgie traditionnelle au Burundi: Techniques et croyances (Archives d'anthropologie, no. 25)*. Tervuren: Musée royal de l'Afrique centrale.
Chikwendu, V. E., Craddock, P. T., Farquhar, R. M., Shaw, T., & Umeji, A. C. (1989). Nigerian sources of copper, lead and tin for the Igbo-Ukwu bronzes. *Archaeometry, 31*(1), 27–36.
Childs, S. T., & Herbert, E. W. (2005). Metallurgy and its consequences. In A. B. Stahl (Ed.), *African archaeology: Acritical introduction* (pp. 276–300). Oxford: Blackwell.
Cline, W. B. (1937). *Mining and metallurgy in Negro Africa* (No. 5). Menasha: George Banta Publishing Company.
Clist, B. (2013). Our iron smelting 14C dates from central Africa: From plain appointment to a full blown relationship. In J. Humpris & Th. Rehren (Eds.), *The world of iron* (pp. 22–28). London: Archetype.
Craddock, P. T. (1995). *Early metal mining and production*. Washington, DC: Smithsonian University Press.
Craddock, P. T. (2000). From hearth to furnace: Evidences for the earliest metal smelting technologies in the Eastern Mediterranean. *Paléorient, 26*(2), 151–165.
Craddock P. T. (2010). New paradigms for old iron: Thoughts on É. Zangato & A.F.C. Holl's "On the Iron Front". *Journal of African Archaeology, 8*, 29–36.
Darling, P. (2013). The world's earliest iron smelting? Its inception, evolution and impact in Northern Nigeria. In J. Humpris & Th. Rehren (Eds.), *The world of iron* (pp. 156–167). London: Archetype.
de Barros, P. (2013) A comparison of early and later Iron Age societies in the Bassar region of Togo. In J. Humpris & Th. Rehren (Eds.), *The World of Iron* (pp. 10-21). London: Archetype.
Edwards, D. N. (2004). *The Nubian past: An archaeology of the Sudan*. London: Routledge.
Emery, W. B. (1963). Egypt exploration society preliminary report on the excavations at Buhen, 1962. *Kush, 11*, 116–120.
Essomba, J.–M. (2004). Status of iron age archaeology in Cameroon. In H. Bocoum (Ed.), *The origins of iron metallurgy in Africa* (pp. 135–148). Paris: ENESCO Publishing.
Eze-Uzomaka, P. (2013). Iron and its influence on the prehistoric site of Leija. In J. Humpris & Th. Rehren (Eds.), *The world of iron* (pp. 3–9). London: Archetype.
Fenn, T. R., Killick, D. J., Chesley, J., Magnavita, S., & Ruiz, J. (2009). Contacts between West Africa and Roman North Africa: Archaeometallurgical results from Kissi, Northeastern Burkina Faso. In S. Magnavita, L. Koté, P. Breunig, & O. A. Idé (Eds.), *Crossroads/Carrefour Sahel: Cultural and technological developments in first millennium BC/AD West Africa* (pp. 119–146). (Journal of African Archaeology Monograph Series 2). Frankfurt a. M.: Africa Magna.
Grébénart, D. N. H. (1987). Characteristics of the final neolithic and metal ages in the region of Agadez (Niger). In A. Close (Ed.), *Prehistory of Arid North Africa: Essays in honour of Fred Wendorf* (pp. 287–316). Dallas: Southern Methodist University Press.
Herbert, E. W. (1984). *Red gold of Africa: Copper in precolonial history and culture*. Madison: University of Wisconsin Press.
Holl, A. (2000). Metals and precolonial African society. In M. Bisson, S. T. Childs, P. de Barros, & A. Holl (Eds.), *Ancient African metallurgy: The sociocultural context* (pp. 1–81). Walnut Creek: AltaMira Press.
Holl, A. F. (2009). Early West African metallurgies: New data and old orthodoxy. *Journal of World Prehistory, 22*(4), 415–438.

Horton, M. (1996). *Shanga: The archaeology of a muslim trading community on the coast of East Africa*. London: British Institute in Eastern Africa Memoir 14.

Huffman, T. N. (2007). *Handbook to the iron age: The archaeology of precolonial farming societies in Southern Africa*. Durban: University of Kwazulu-Natal Press.

Humpris, J. and Iles, L. (2013). Pre-colonial iron production in Great Lakes Africa: recent research at UCL Institute of Archaeology. In J. Humpris & Th. Rehren (Eds.). *The World of Iron* (pp. 56–64). London: Archetype.

Killick, D. J. (1987). On the dating of African metallurgical sites. *Nyame Akuma, 28,* 29–30.

Killick, D. (2004a). Review essay: What do we know about African iron working? *Journal of African African Archaeology, 2*(1), 97–112.

Killick, D. (2004b). Social constructionist approaches to the study of technology. *World Archaeology, 36*(4), 571–578.

Killick, D., Van der Merwe, N. J., Gordon, R. B., & Grébénart, D. (1988). Reassessment of the evidence for early metallurgy in Niger, West Africa. *Journal of Archaeological Science, 15*(4), 367–394.

Kiriama, H. O. (1987). Archaeo-metallurgy of iron smelting slags from a Mwitu Tradition site in Kenya. *The South African Archaeological Bulletin, 42,* 125–130.

Klemm, D., Klemm, R., & Murr, A. (2001). Gold of the Pharaohs–6000 years of gold mining in Egypt and Nubia. *Journal of African Earth Sciences, 33*(3), 643–659.

Lankton, J., Ige, A., & Rehren, T. (2006). Early primary glass production in southern Nigeria. *Journal of African Archaeology, 4,* 111–138.

Levtzion, N. (1973). *Ancient Ghana and Mali* (Vol. 7). London: Methuen.

Mapunda, B. B. (1997). Patching up evidence for ironworking in the Horn. *African Archaeological Review, 14*(2), 107–124.

Martinón-Torres, M., & Rehren, Th. (2009): Post-medieval crucible production and distribution: a study of materials and materialities. *Archaeometry, 51,* 49–74.

Martinón-Torres, M., Rehren, T., & Freestone, I. C. (2006). Mullite and the mystery of Hessian wares. *Nature, 444*(7118), 437–438.

McIntosh, S. K., & McIntosh, R. J. (1988). From stone to metal: New perspectives on the later prehistory of West Africa. *Journal of World Prehistory, 2*(1), 89–133.

Midant-Renes, B. (2000). The Naqada period (c. 4000–3200 BC). In I. Shaw (Ed.), *The Oxford history of ancient Egypt* (pp. 44–60). Oxford: Oxford University Press.

Miller, D. E., & Van der Merwe, N. J. (1994). Early metal working in sub-Saharan Africa: A review of recent research. *Journal of African History, 35*(1), 1–36.

Okafor, E. (1993). New evidence on early iron-smelting in Southeastern Nigeria. In T. Shaw, P. Sinclair, B. Andah, & A. Okpoko (Eds.), *The archaeology of Africa: Food, metals, and towns* (pp. 432–448). London: Routledge.

Pernicka, E., Begemann, F., Schmitt-Strecker, S., Todorova, H., Kuleff, I. (1997). Prehistoric copper in Bulgaria. Its composition and provenance. *Eurasia Antiqua, 3,* 41–180.

Person, A & Quechon, G. (2004). Chronometric and chronological data on metallurgy at Termit: Graphs for the study of the ancient iron ages. In H. Bocoum (Ed.), *The origins of iron metallurgy in Africa*. (pp. 119–126). Paris: UNESCO Publishing.

Phillipson, D. W. (2005). *African archaeology*. Cambridge: Cambridge University Press.

Pikirayi, I. (2001). *The Zimbabwe culture: Origins and decline in Southern Zambezian states* (Vol. 3). Walnut Creek: AltaMira.

Pieterse, J. N. (1998). *White on black: images of white and blacks in western popular culture*. New Haven: Yale University Press.

Pwiti, G. (1996). *Continuity and change: An archaeological study of farming communities in Northern Zimbabwe AD 500–1700*. Uppsala: Societa Archaeologica Uppsaliensis.

Radivojević, M., Rehren, T., Pernicka, E., Šljivar, D., Brauns, M., & Borić, D. (2010). On the origins of extractive metallurgy: New evidence from Europe. *Journal of Archaeological Science, 37*(11), 2775–2787.

Rehren, Th. (2001). Meroe, iron and Africa. *Der Antike Sudan, 12,* 102–109.

References

Rehren, T., Belgya, T., Jambon, A., Káli, G., Kasztovszky, Z., Kis, Z., & Szőkefalvi-Nagy, Z. (2013). 5000 years old Egyptian iron beads made from hammered meteoritic iron. *Journal of Archaeological Science, 40*(12), 4785–4792.

Reimer, P. J., Bard, E., Bayliss, A., Beck, J. W., Blackwell, P. G., Ramsey, C. B., ... & van der Plicht, J. (2013). IntCal13 and Marine13 radiocarbon age calibration curves 0–50,000 years cal BP. *Radiocarbon, 55*(4), 1869–1887.

Scheel, B. (1989). *Egyptian metalworking and tools.* Oxford: Shire Publications.

Schmidt, P. R. (1997). *Iron technology in East Africa: Symbolism, science, and archaeology.* Bloomington: Indiana University Press.

Shaw, T. (1970). *Igbo-Ukwu: An account of archaeological discoveries in Eastern Nigeria* (Vol. 2). Evanston: Northwestern University Press.

Shinnie, P. L. (1985). Iron working at Meroe. In R. Haaland & P. L. Shinnie (Eds.), *African iron working-ancient and traditional* (pp. 28–35). Oslo: Norwegian University Press.

Stahl, A. B. (2014a). Africa in the World: (Re)centering African history through archaeology. *Journal of Anthropological Research, 70*(1), 5–33.

Summers, R. (1969). *Ancient mining in Rhodesia and adjacent areas.* Salisbury: Trustees of the National Museums of Rhodesia.

Thomas, N. (1991). *Entangled objects: exchange, material culture and colonialism in the Pacific.* Cambridge: Harvard University Press.

Thornton, C. P., & Rehren, T. (2009). A truly refractory crucible from fourth millennium Tepe Hissar, Northeast Iran. *Journal of Archaeological Science, 36*(12), 2700–2712.

Tylecote, R. F. (1975). The origin of iron smelting in Africa. *West African Journal of Archaeology, 5,* 1–3.

Tylecote, R. F. (1976). *A history of metallurgy.* London: Metals Society.

van Grunderbeek, M.-C., Roche, E., & Doutrelepont, H. (2001) Un type de fourneau de fonte de fer associé à la culture Urewe (Aˆ ge du Fer Ancien) au Rwanda et au Burundi. *Mediterranean Archaeology, 14,* 291–297.

Van Der Merwe, N. J., & Avery, D. H. (1982). Pathways to Steel: Three different methods of making steel from iron were developed by ancient peoples of the Mediterranean, China, and Africa. *American Scientist,* 146-155.

Wilson, A. (2012). Saharan trade in the Roman period: Short-, medium-and long-distance trade networks. *Azania: Archaeological Research in Africa, 47*(4), 409–449.

Zangato, E., & Holl, A. F. (2010). On the iron front: New evidence from North-Central Africa. *Journal of African Archaeology, 8*(1), 7–23.

Chapter 3
Mother Earth Provides: Mining and Crossing the Boundary Between Nature and Culture

Introduction

Ores are rocks whose metal content is economically exploitable using available technology (Chirikure 2010a). Ore deposits are often formed where subsurface geological processes have removed metals from common rock or from masses of molten magma, and have redeposited them in other locations at much higher concentrations (Killick 2014b) (Fig. 3.1). Alternatively, ore deposits are formed where surface processes have eroded minerals from rocks and concentrated them elsewhere. These processes created two major ore forms: epigenetic and syngenetic. Epigenetic ores were introduced into their surrounding rocks after the host rock had already formed. Usually, such ores mineralized in the form of lodes and veins. In contrast, syngenetic deposits were formed at the same time as the host rock. The erosion of the two types of ore bodies through fluvial processes often transports small particles of ore which when found in sufficient concentration may be mined as alluvial or eluvial deposits.

Because ores abound in nature, the process of ore procurement in Africa involved negotiations between the living, the dead and the deities through the mediation of intermediaries such as spirits of the land. For the living to cross the nature–culture boundary to extract ores from the earth's belly, a number of rituals and taboos were conducted to propitiate ancestors. There are few archaeological traces of rituals associated with mining. However, ethnographically, in most parts of sub-Saharan Africa from the Dogon of Mali in West Africa to the Njanja of Zimbabwe, miners used medicines to neutralize malevolent influences during the process of mining (Chirikure 2006; Huysecom and Augustoni 1997). The concept of pollution often expressed through forbidding menstruating women from mines was important in sub-Saharan mining (Cline 1937; Haaland 2004a; Herbert 1993; Schmidt 2009). However, as with everything else, the diversity of African practice makes it dangerous to generalize ethnographically and archaeologically because women worked open pit copper mines in Katanga, Democratic Republic of Congo, in the historical and ethnographic periods, and archaeologically, female skeletons have been found

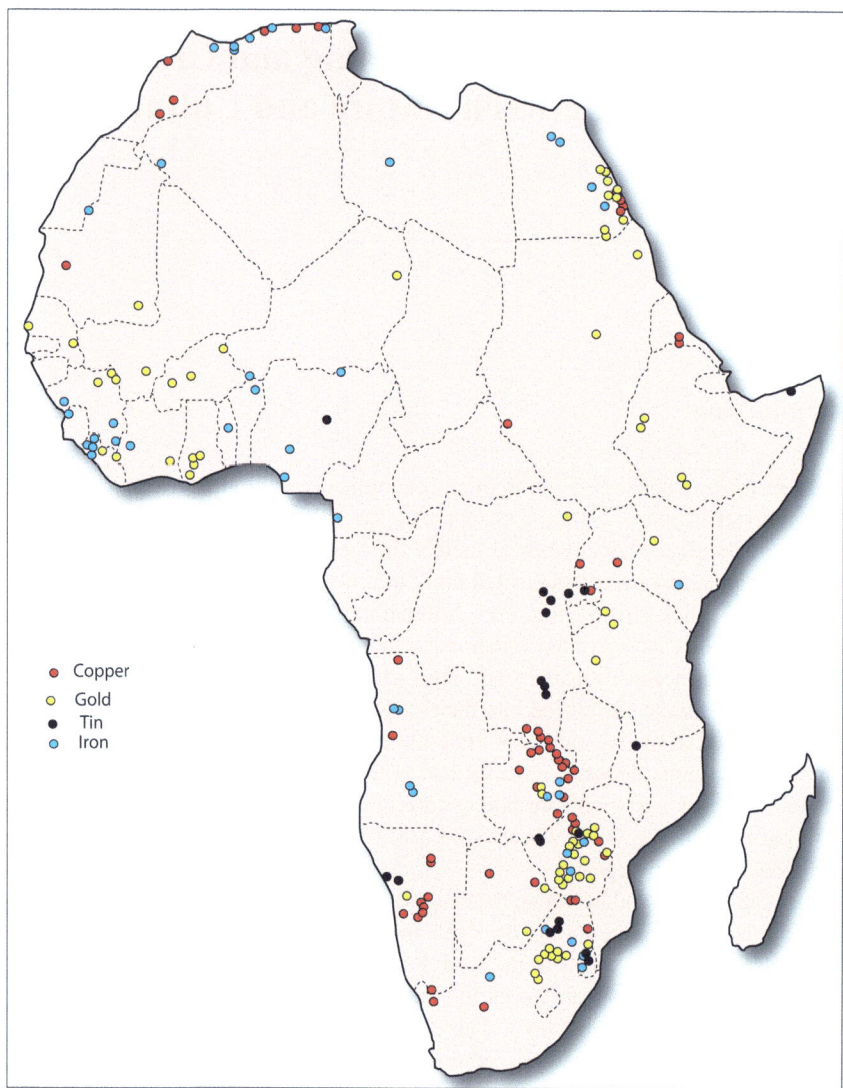

Fig. 3.1 Location of mines worked in African antiquity. Note that because of its abundance, iron was worked in more areas than illustrated here

in collapsed gold mines dating to the mid-second millennium AD in Zimbabwe (Summers 1969).

In cases where miners failed to obtain suitable ores, sacrifices were also offered, resulting in the deities unlocking the earth by providing ore (Haaland 2004b; Huysecom and Augustoni 1997). It was the ancestors and deities who had the power to mediate between nature and culture and between the known and unknown (Mbiti 1990). The intervention of deities in mining as well as the miners' belief in supernatural powers is not unique to Africa. In current-day Bolivia, El Tio, a devil figure has power over the mines and miners and is believed to have simultaneously offered protection and destruction of miners since pre-Columbian times (Nash 1993). At Timna in Israel, copper miners invoked the power of gods and the supernatural as evidenced by the presence of an Egyptian temple which serviced miners at the mining site (Rothenberg 1999; Rothenberg and Bachmann 1988).

Armed with intent to produce metal, and the power of supernatural forces, prehistoric miners proceeded to extract the ore from the ground (Childs 1998). Most of the time, miners were smelters who knew the desired quality of the ore. In Africa, as elsewhere, there exist many metallogenic provinces and geological bodies rich in exploitable iron, copper, tin and gold (Robb 2009). Examples of these include the Lubumbashi (formerly Katanga) region in the Democratic Republic of Congo, rich in copper and iron; the South African Phalaborwa Carbonatite Complex, rich in iron and copper ores; and the Bambuk goldfields of West Africa situated on the upper Senegal River area (Bisson 2000; Curtin 1973; Phimister 1974; Robb 2009; see Fig. 3.1). In Africa, archaeological signatures dating from as early as the Early Iron Age indicate that these types of ore bodies were exploited for almost 2000 years before the onset of colonialism in the late nineteenth century (Summers 1969). The mining techniques were not static, and neither was the scale and organization of production throughout antiquity. In some cases, free labour was used, but slavery was also a source of labour in ancient Egypt and the Roman times. As such, the beliefs and values that were prevalent in society were also produced and reproduced during mining.

Different methods of ore extraction were adopted, depending on the nature of the mineralization. In regions where natural processes of erosion resulted in the deposition in downstream areas of rich but fine grains of ores, alluvial mining was practised, as in the Egypt Eastern Desert around 2000 BC. The Mafa of Northern Cameroon (David et al. 1989) and the Kikuyu (Brown 1995) of Kenya panned magnetite sand for bloomery iron smelting in the more recent past. Similarly, the Soninke people of ancient Ghana panned gold from the rich goldfields of Bambuk between cal AD800 and 1000 (Levtzion 1973). In areas where rich outcrops of ore existed such as at Thabazimbi, the iron mountain of South Africa, no excavation was required, resulting in mere surface collection of ores (Friede and Steel 1976). As surficial resources were depleted, miners would follow the lodes horizontally or vertically, resulting in either open or underground mining. Geological constraints such as the nature of the mineralization motivated for the use of similar mining techniques in both the Old and New Worlds, be it at Potosi, Rio Tinto, Timna, Phalaborwa or Zawar (Chirikure 2010a; Hammel et al. 2000; Hauptmann 2007; Summers 1969). Importantly, not all 'ore bodies' were worked due to limitations posed by the available extractive metallurgy. For example, sulphidic copper ores were avoided by smelters in much of sub-Saharan Africa because the widely available single-stage smelting process was not optimized to deal with them (Miller and Killick 2004).

In sum, the process of finding ores and extracting them from the ground as documented in the historical period in Africa and Latin America and in antiquity at Timna in the Arabah Desert required the intervention of ancestors and deities as well as an understanding of the technical knowledge of the local geology and the behaviour of the ore body. This intersection between technology and culture made preindustrial mining across the globe a socially embedded process.

History of Mining: A Global Perspective

The point that preindustrial mining was not static throughout antiquity needs no emphasis. However, where and when did mining begin in the world? This broad question cannot be answered with certitude but it is now a well-established fact that during the Middle Stone Age between 200,000 and 40,000 years ago, early humans residing in Africa extracted and modified iron oxides for use as pigments (Watts 2002). The Ngwenya mines on the Bomvu Ridge in Swaziland which were dated by radiocarbon to 40,000 BP are further testimony to the antiquity of mining in Southern Africa (Dart and Beaumont 1967). However, the earliest intentional mining of oxide and carbonate ores to gain metal through a controlled application of heat did not occur in Africa (Craddock 2000). The available evidence suggests that around 5000 BC, communities living in the Middle East and the neighbouring Balkans were smelting copper ores to produce metal for tool making. Mines such as Rudna Glava and Ai Bunar in Bulgaria and Serbia were worked during this period (Radivojević et al. 2010). Initially, while ores were possibly surface-collected, as outcropping deposits got depleted and populations increased, more complex methods of mining developed, resulting in open and underground mining at places such as Faynan in Jordan, Timna in Israel, Laurion in Greece and Rio Tinto in Spain (Hauptmann 2007).

In Africa, the available evidence indicates that mining with intent to obtain ores for metallurgical reasons was established by the fourth millennium BC in Egypt (Klemm and Klemm 2012) and by first millennium BC in regions such as West Africa (Holl 2009; Killick et al. 1988). In Southern Africa, the southward migration of the Bantu introduced metallurgically motivated mining to this region early in the first millennium AD. In these regions, the nature of the geology too resulted in humanity responding to the need to extract ore in identical and varying ways.

As elsewhere in the Old World, ancient mining was a key component of precolonial Africa's social, technical and economic systems. It became the pivot on which long-distance trade, local trade and the rise and fall of empires were anchored (Chirikure 2007; Phimister 1974). For example, the depletion of gold fields under its control precipitated the decline of ancient Ghana at the onset of the second millennium AD, while at the same time promoting the fortunes of Mali (Levtzion 1973). Similarly, the rise of lucrative gold mining in Northern Zimbabwe has been strongly implicated in the decline of Great Zimbabwe (Pikirayi 2001). The most pertinent observation, however, is that unlike some twentieth-century industrial mines across Africa, which became centres of urbanization, in precolonial Africa as elsewhere, early mining landscapes were not always associated with urban settlements (Mackenzie 1975; Summers 1969; Van der Merwe and Scully 1971). It seems that at places such as Timna and Faynan, metal workers were associated with village-scale settlement but some settlements developed over time as the scale of production increased (Hauptmann 2007). The absence of rich ore sources near the Early Iron Age site of Swart Village (cal 700 to 1200 AD) near Mt Darwin in Northern Zimbabwe suggests that smelters at the site brought ore from elsewhere (Chirikure and Rehren 2006). With time, large-scale smelting took place at sources, resulting in large mounds of slag around Phalaborwa in South Africa, Meroe in the Sudan, the Dogon of Mali (>300,000 m^3 of slag) (Robion Brunner et al 2013), Bassar in Togo (>80,000 m^3 of slag) (de Barros 2013), at Laurion in Greece, in Cyprus and the Alps region in Austria. According to Summers (1969), most modern mines in Africa as elsewhere in the world are situated on sites of preindustrial mining, showing continuity in humanity's dependence on minerals since metallurgy began. This overlay of later evidence on earlier traces has destroyed valuable early evidence.

The First Step: Ore Exploration and Prospecting in Precolonial Africa

According to Cline (1937), most well-known ore deposits across much of sub-Saharan Africa were worked for successive millennia. Iron, gold, tin and copper are some of the metals whose ores were continuously worked preindustrially. Mining landscapes such as Katanga in the Democratic Republic of Congo, the Copper belt in Zambia, Bambuk on the upper Senegal River, the historic Gold Coast, the Zimbabwe plateau and Phalaborwa are but few examples of metallogenic provinces that best fit this description. For example, the copper province of Central Africa (Katanga and Copperbelt of Zambia) was worked from c.AD200–1900 (Bisson 2000) while the gold deposits on the Zimbabwe plateau were worked between c.AD1000 and 1900. To identify mineralization, preindustrial Africans would have prospected and explored the land. In geological terms, mineral prospecting is the process of exploring the landscape in search of exploitable mineral deposits (Robb 2009). Knowledge of the earth's physical landscape was important to the prospectors because different metals are geologically specific and are associated with different types of host rocks and vegetation regimes (Summers 1969). In some cases, outcropping ores may have provided some additional leverage. Without doubt, it is clear that these extinct prospectors were advanced in their reading of the landscape to distinguish ore-bearing from non-ore-bearing rocks. The colour of the ore-bearing rocks would certainly have helped (Hauptmann 2007). Successful exploration would also have demanded complex methods of organization and mobilization of labour. In ethnographically known examples, sizeable groups of people scoured the landscape for viable ore sources and the head smelters certified the quality of the ore through visual inspection (Childs 1998; Mackenzie 1975). Once identified, certain ore bodies were worked successively over millennia.

Prospecting too was guided by the intervention of ancestors in the ethnographic record (Childs 1998) such that amongst the Tsara of Ethiopia, the discovery of a mine was followed by a series of rituals including pouring beer on the ground to acknowledge the role of ancestors in guiding the prospectors (Haaland 2004b). Conversely, failure to find a rich ore deposit prompted the Dogon prospectors of Mali to slaughter animals for ritual cleansing and to look for medicines to chase away bad luck and evil spirits. It is possible that preindustrial prospecting was associated with a range of rituals and beliefs pertinent in contexts of success just as they were in failure. Whereas an absence of written documents and a lack of tangible evidence makes it difficult to reconstruct the nature of prospecting and exploration in much of preindustrial Africa, it was nonetheless a key component of the *chaîne opératoire* of metal working.

Methods of Crossing the Nature–Culture Boundary: Mining of Ores in Preindustrial Africa

Once ore bodies were located, numerous techniques were used to dig up the ore, and these were largely motivated by the nature of the mineralization and the opportunities and or constraints that it imposed. For example, alluvial and lode deposits demanded not just varying techniques of mining, but also different tools used in separating the ore from the host rock or material. Within variation, four major types of ore mining were used across the world before industrialization (Chirikure 2010a; Craddock 1995). These are surface collection, alluvial mining, open mining and underground mining. It would seem that with very rich deposits, miners often started working outcropping ore and followed the mineralization into the ground once surficial deposit was exhausted (Hammel et al. 2000). The nature of the lode determined whether open or underground methods were to be used. Alluvial methods were also used to mine placer deposits, usually after the rainy seasons (Curtin 1973; Phimister 1974). On a global level, the methods for mining in the preindustrial world were context dependent but almost universal, constrained by the available techniques, the underlying geology and the prevailing sociocultural context.

Surface Collection

In areas with rich ore mineralization, there was no need to dig deep into the ground to extract suitable ore. Instead, the miners simply surface collected high grade ores for smelting. Craddock (2000) argues that the earliest smelted copper ores in Egypt c. 4000 BC may have been very pure because they left little slag and may have been surface-collected. There is no tangible evidence for surface collection in the archaeological record beyond this inference. Even in Africa's recent past, there are numerous examples where miners collected rich nodules of iron ore and smelted them to gain elemental metal. Across the continent, the surface collection of iron seems to have been a very popular activity (Cline 1937). For example, nineteenth-century miners in Nyanga, Northeastern Zimbabwe surface collected lateritic ores and hematite for smelting (Chirikure and Rehren 2004). At Musina in modern-day South Africa, the Venda and Lemba people also surface-collected rich iron ores and smelted them to produce metal (Mamadi 1940). Surface collection was practised in numerous metallogenic provinces of sub-Saharan Africa such as Katanga and Bassar where, respectively, copper and iron exist in abundance (Cline 1937; de Barros 1988). Surface collection did not leave much in the form of traces in antiquity but it is tempting to speculate that it was one of the earliest methods of accessing ore and that with increased scale of production other methods came into play. Continual surface collection, however, exhausted outcropping deposits, which motivated finding methods that required excavation into the earth's crust and the accompanying but varying intensities of earth moving.

Alluvial Mining

Alluvial mining was one of the most common methods of ore extraction in preindustrial Africa, and indeed in the whole world (Summers 1969). Alluvial deposits form along waterways and rivers, when small quantities of ore are washed from source and deposited on river beds and gravels (Robb 2009). Normally, this took place in cases where rivers flow near or across auriferous and other ore-rich outcrops. There are a few metallogenic provinces that since time immemorial were the center of alluvial mining in South Central and West Africa (Curtin 1973; Phimister 1974) (Fig. 3.1). Examples of these include Bambuk on the upper Senegal River, Mandara Mountains of Cameroon, Northern Zimbabwe and tons of magnetite scree (eluvial) around the former Lolwe Hill at Phalaborwa which may have been exploited since AD1000 (Killick and Miller 2014).

Ethnographic and Historical Descriptions of Alluvial Mining The ethnographic examples collated by Cline (1937) indicated that gold, tin and iron were the only metal ores that were extracted from alluvial ore in precolonial Africa. Alluvial deposits were worked using a process known as panning which involved digging ore-rich sand on the river beds and scooping it into hemispherical buckets or bowls for density separation through shaking or winnowing (Chirikure 2010a). During this process, the lighter sand matrix stayed on top and was thrown away while the heavy metal settled at the bottom. Across many areas of Africa, runoff from heavy rains often carried with it small magnetite particles that were deposited along river banks and other erosion channels. In regions such as Northern Cameroon, Kenya, and Southwestern Nigeria, these magnetite sands were collected in exploitable quantities. Not surprisingly, the Mafa of Northern Cameroon, the Yoruba of Southwestern Nigeria and the Kikuyu of Kenya panned rich magnetite sands which they smelted to produce a mix of cast and soft iron in the recent past (Brown 1995; Cline 1937; David et al. 1989; Ige and Rehren 2003). Cline (1937) vividly described how Kikuyu men and women diverted water into small artificial lakes to wash the magnetite-rich sands, thereby separating the ore from the surrounding sandy matrix.

Fig. 3.2 Akan gold miners diving into the Ankobra River to extract diamond-rich sand which was panned on the river bank.. (Redrawn from Garrard 2011, p. 116, original from Dapper 1668)

Alluvial gold mining was recorded in detail by the Portuguese who operated in Northern Zimbabwe from the sixteenth century AD onwards. In this region, the Shona people worked the alluvial deposits along rivers such as Mukaradzi and Mazowe in Northern Zimbabwe (Phimister 1974; Swan 1994). Ellert (1992) describes a very complicated method of working alluvial gold deposits used by the Shona in Northern Zimbabwe in the nineteenth and twentieth centuries. This is the process of underwater dredging whereby experienced divers draped weights on their backs and dived into the waters of the Mazowe and Rwenya rivers. Here, they scooped auriferous sand into receptacles and took it to the surface where women and children separated the gold from the sand (Chirikure 2010a). Interestingly, the Dutch painter Olfert Dapper painted a seventeenth-century Akan community dredging auriferous sand from the Ankobra River in Southern Ghana, suggesting that the technique may have some wide expression throughout Africa (Fig. 3.2) (Garrard 2011, p. 116).

It seems that alluvial mining was practised by both men and women. In particular, men dug the earth while women winnowed, but this division of labour varied from context to context. It should be pointed out that there are rare instances where alluvial miners dug wide and deep holes into the sand, resembling open mining (Cline 1937). The ore–debris separation was similarly carried out by women using density separation techniques such as winnowing. It has been argued that women featured prominently in panning because the principle of the process was rooted in winnowing, an occupation gendered as female across many African societies (Summers 1969). This is one remarkable example of crossovers in gender roles which, as we have seen in the extraction of magnetite sands by Kikuyu women, deviates from generalizations which see no space and place for women in African mining.

Archaeological Insights into Alluvial Mining Alluvial mining is difficult to explore archaeologically because traces of mining were often overwritten after the following year's rains. In an archaeologically rare case, Wagner and Gordon (1929) recorded undated late second millennium AD finds of alluvial tin ore at a few tin-smelting archaeological locations in the Blaauwbank Donga near Rooiberg in the Southern Waterberg of South Africa (Fig. 3.1). In most cases, what we know about alluvial mining in African archaeology is derived either from written sources or from secondary sources such as the geochemistry of objects and slags. In Southern Africa, the geochemistry of tin slags dating from AD1600 revealed that some slags—particularly those from the Blaauwbank Donga—were rich in zircon, ilmenite and other heavy detrital minerals (Miller and Hall 2008). This contrasted with the composition and mineralogy of slags from Smelterskop which lacked detrital minerals. Chirikure et al. (2010) concluded on the basis of these differences that Rooiberg tin miners exploited both alluvial and hydrothermal tin.

Arabic writers such as Ibn Battuta indicate that the fabled Bambuk gold fields in West Africa were worked alluvially during the time of the Soninke Empire of ancient Ghana in the late first to early second millennium AD. Later its successor empires Mali and Songhai tapped into the gravels of the upper Senegal River and adjacent tributaries, extracting tons of ore which ended up in the Islamic world via the trans-Saharan trade (Curtin 1973). In southern Africa, Al Masudi noted that the inhabitants of Sofala widely believed to be the Zimbabwe plateau exploited alluvial gold in the early second millennium AD. As indicated in textual evidence, alluvial mining was one of the methods practised by ancient Egyptians and Nubians from 4000 BC onwards (Klemm and Klemm 2012).

Summary Hammel et al. (2000) contend that alluvial mining was a fairly simple method of ore extraction, despite the fact that it required complex methods of decision-making and organization. For instance, in the ethnographic record, panners had to understand rainfall cycles as well as to separate rich deposits from poor ones so as to make their labour input worthwhile. Furthermore, the miners had to carefully schedule their activities depending on seasons to free labour for critical pursuits such as agriculture in the rainy season (Curtin 1973; Phimister 1974; Summers 1969). Consequently, most panning activities—whether associated with magnetite sands or gold extraction—were confined to the winter months when the water table was low and communities had just finished harvesting. Annually, the end of alluvial mining cycle coincided with the beginning of the rainfall season when labour was reallocated to pursuits such as farming and cattle herding.

Open Mining

Across the African continent, the nature of the ore mineralization mandated that trenches had to be excavated into the ground to extract ores. According to Summers (1969), once surface outcropping ore was depleted, miners followed the deposit into the ground. If the mineralization was horizontal, long trenches were dug, but if the mineralization was vertical, miners would dig deeper creating open pits of varying sizes. In contrast to underground mines explained below, open mines are more horizontal and could reach a distance of more than 380 m (Bisson 2000). Copper, tin, gold and iron were all extracted using the technique of stope mining. The process of stope mining involved excavating ore mineralization from the ground using basic tools such as hoes, shovels and picks. The ore was separated from the host rock, resulting in the creation of mining dumps of varying sizes on the sides of the mines. Most of these preindustrial mines were not backfilled; because of this, significant amounts of tangible evidence littered different parts of Africa before the onset of modern mining (Cline 1937; Summers 1969). In cases where miners encountered very hard rocks, difficult to break with the limited tools in their arsenal, they fired those rocks to very high temperatures and poured water on them while they were still hot leading to a cracking of the rock which can then be mined more easily. This technique known as 'fire setting' worked remarkably well and conferred the added advantage to archaeologists in that it left charcoal which can be radiometrically dated to estimate the age of the mine or the period when the mine was worked.

Ethnographic and Historical Descriptions One of the most detailed cases of open mining relates to copper mining by the Yeke of the Democratic Republic of Congo and the Kaonde of Zambia (Fig. 3.1). According to Cline (1937), in the nineteenth and early twentieth centuries, the renowned copper workers such as the Yeke and Kaonde worked open copper mines, digging into the ground to extract the ore from the lodes. These copper miners of Central Africa also practised fire setting to break hard host rocks at mines such as Etoile and Dikuluwe (Bisson 2000). The bigger lumps were broken by hammerstones into smaller pieces which were easily transportable to the surface. Interestingly, one of the expeditions was led by a woman (Bisson 2000).

Mackenzie (1975) discusses open mining for banded iron stone on the historically famous Hwedza range of Mountains in the eighteenth and nineteenth centuries which rarely exceeded 4 m in diameter. Baskets were used to hoist the ore and associated rocks. Where possible, the ore was separated from the host rock inside the mine. In and around Oyo in Nigeria, a number of long trenches representing open mines have also been reported (Bellamy and Habord 1904) (Fig. 3.1).

Archaeological Descriptions One of the most detailed and vivid descriptions of open mining in precolonial Africa is provided by Summers (1969) who studied in detail the ancient copper and gold workings found in Zimbabwe and adjacent territories dating between the late first and mid-second millennium AD. Summers explains that the epigenetic quartz vein ore mineralization presented notable constraints that shaped open gold mining on the Zimbabwe plateau from the late first millennium AD onwards. For instance, the small width of the lodes and veins dictated that most gold mines were only a few metres in diameter. Such a width was also determined by the need to create a comfortable working space for the miners. Occasionally, when two gold reefs were adjacent to each other, and were detached by soft rock, they were mined simultaneously using the technique of side stoping. At Thakadu in North-Eastern Botswana, the technique of open mining was widely used to extract copper from AD1450 onwards (Huffman et al. 1995).

The available evidence indicates that women worked some of the open mines. For example, amongst the Shona people of Zimbabwe, women worked inside gold mines. Similarly, women also worked the Katanga copper mines in central Africa. At Kansanshi in Zambia, copper was mined using open mining methods from AD 300 onwards (Bisson 2000). The earliest example of open mining in Africa comes from Egypt and Nubia and is associated with gold extraction (Klemm and Klemm 2012). There are a number of gold mines in the Eastern Desert which represent abandoned open but not very deep shafts worked from dynastic periods to later times. Hammerstones and ore milling infrastructure was also documented here. In sub-Saharan Africa, the fact that most mines were worked over generations resulted in the obliteration of earlier evidence by most recent mining. Summers (1969) thinks that some open gold mines on the Zimbabwe plateau date to the late first millennium AD. Superficially, this may appear to contradict the earliest finds of gold which date to the early second millennium AD. On closer observation, this may be true given that glass beads from the East African coast start appearing in Southern Africa from AD700 onwards. Arabic writers mention that gold was one of the commodities sourced in Southern Africa. The only controversial and wildly speculative conclusion from Summers (1969) is the innuendo that Indians worked the gold mines on the Zimbabwe plateau. This is invalidated by skeletal evidence which shows that local people worked the mines. In any case, the nature of the mineralization dictated that miners had to dig into the ground resulting in similar techniques across the world throughout antiquity. There is simply no other way to extract subsurface ore other than digging.

Underground Mining

The technique of open mining was not well suited to work horizontal lodes which lay at some depth underground. To exploit these, precolonial peoples first made vertical shafts into the ground and mined out those underground reefs, giving rise to underground mining. During preindustrial times, it appears that different smelting communities across the world—be it at Timna in Israel (Rothenberg 1962), Zawar in India (Willies et al. 1984), Bambuk in West Africa or the Late Bronze Age mines on the Austrian Alps—developed underground mining when the open mines reached great depths. The underground deposits were mined out using procedures identical to those employed during open mining. The only difference is that underground, ore mineralization often branched into various directions which, when mined, created a network of tunnels, galleries, adits and mined out pockets beneath the ground. As with open mining, copper, tin, iron and gold were extracted using the technique of underground mining. However, there are very few instances in precolonial Africa where miners created underground tunnels and galleries during iron mining. This probably stems out of the metal's relative abundance on the earth's crust.

Ethnographic and Historical Descriptions Cline's (1937) compilation of ethnographic cases of mining in sub-Saharan Africa exposed that underground mining was practised by many groups to work gold, copper and possibly iron and tin. The Venda and Lemba of Musina extracted carbonate copper ores from the copper mines around Musina in Northern South Africa in the nineteenth century (Stayt 1931). These mines have since been destroyed by modern mining. Similarly, the Ba-Phalaborwa of South Africa people also extracted copper ores from the underground mines on Lolwe Hill. One of the largest copper mines in precolonial Africa, Etoile, near modern-day Lubumbashi in the Democratic Republic of Congo, was also mined using underground methods in the late nineteenth and early twentieth centuries (Bisson 2000).

Archaeological Descriptions Predevelopment impact assessments carried out before the onset of modern mining at Harmony Block in Northeastern South Africa and at Lolwe in Phalaborwa resulted in some of the most detailed archaeological descriptions of precolonial underground mining (Fig. 2.1). Amongst the most dramatic examples of preindustrial underground mines in Southern Africa is the cal thirteenth-century AD copper mine previously located on the Harmony block in the North-Eastern Limpopo Province of South Africa (Evers and van den Berg 1974). In the 1970s, an interdisciplinary team of archaeologists and geologists led by Evers carried out salvage work in advance of modern copper mining. The rescue team documented in remarkable detail the evidence from the mine. Research indicated that the Harmony copper mine consisted of one open stope which branched into at least 25 shafts, pockets and galleries underground. These galleries provided deep insights into indigenous people's understanding of structural geology, mining and the need for ventilation. In Unit 22, Evers and van den Berg (1974) discovered sizeable quantities of hardwoods, probably used as ladders for accessing the mine and or for hoisting ore to the surface. Amazingly, three wooden props used to provide a structural support to forestall collapse were still in place. In another section of the mine, the miners strategically left blocks of unmined rock as pillars to further provide structural support. These techniques are akin to those used in contemporary mining and demonstrate how advanced these preindustrial miners were.

This practice of leaving unmined host rock as pillars underground to provide support was also noted at the Aboyne Gold Mine in Zimbabwe showing that it was widely used in Southern Africa (Summers 1969) (Fig. 3.3). However, such interventions often fell short. The Aboyne Mine collapsed and in the process killed at least four individuals, some of them women. Presumably, the dangers associated with this activity required the intervention of supernatural powers. This is why ancestors and the supernatural seem to have played a significant role in mining in preindustrial Africa and elsewhere (see Nash 1993 for Bolivia).

Underground Mining 47

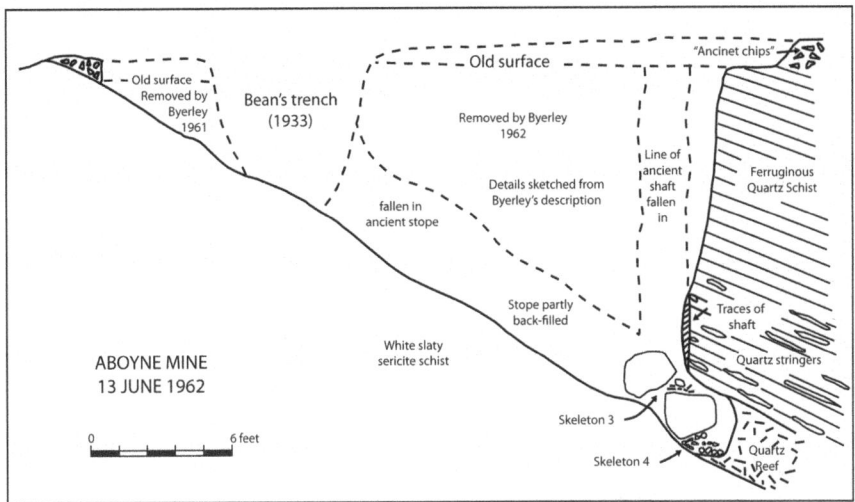

Fig. 3.3 Cross section of the Aboyne gold mine, Central Zimbabwe. (Redrawn from Summers 1969, p. 24, Fig. 5)

The Rooiberg tin mines, situated in the Thabazimbi district of Limpopo Province, South Africa, provides another graphic example of underground mining. The mines were worked between the fifteenth and nineteenth centuries (Chirikure et al. 2007; Hall 1981), but new research has extended the beginning to the thirteenth century AD (Bandama 2013). According to Baumann (1919), the Rooiberg tin mines were established by digging vertical and inclined shafts into the earth's crust. While underground, the tin lodes were followed by drives to pockets (Trevor 1906). It has been estimated that about 18,000 t of ore were mined by indigenous peoples before the onset of modern mining. Interestingly, some of the underground adits at Rooiberg were very narrow, implicating the use of child labour and or very uncomfortable working conditions for adults (Friede and Steel 1976). As with other African underground mines, backfilling was done primarily to reduce the volume to be hoisted, provide stability to the underground building and manage air flow in order to close off 'dead' parts of the mine (Fig. 3.4).

Ladders may have been used for access (Hall 1981). The ore-rich rock was sometimes processed underground using grinding stones similar to those used for processing grain. Examples of these were recovered inside the Rooiberg mines. In terms of division of labour, this may suggest the involvement of women since grinding stones were mainly associated with them just as winnowing and alluvial mining. The ore was likely transported to the top using baskets, while the gangue material was used as backfill. Because underground miners sometimes encountered poisonous gases, some of which were life-threatening, backfilling was one of the interventions made secondarily to neutralize their harmful effects (Hammel et al. 2000). To improve ventilation, the miners at Rooiberg dug narrow and vertical shafts from the ground into the mining chambers. It is possible that a fire was lit underground to

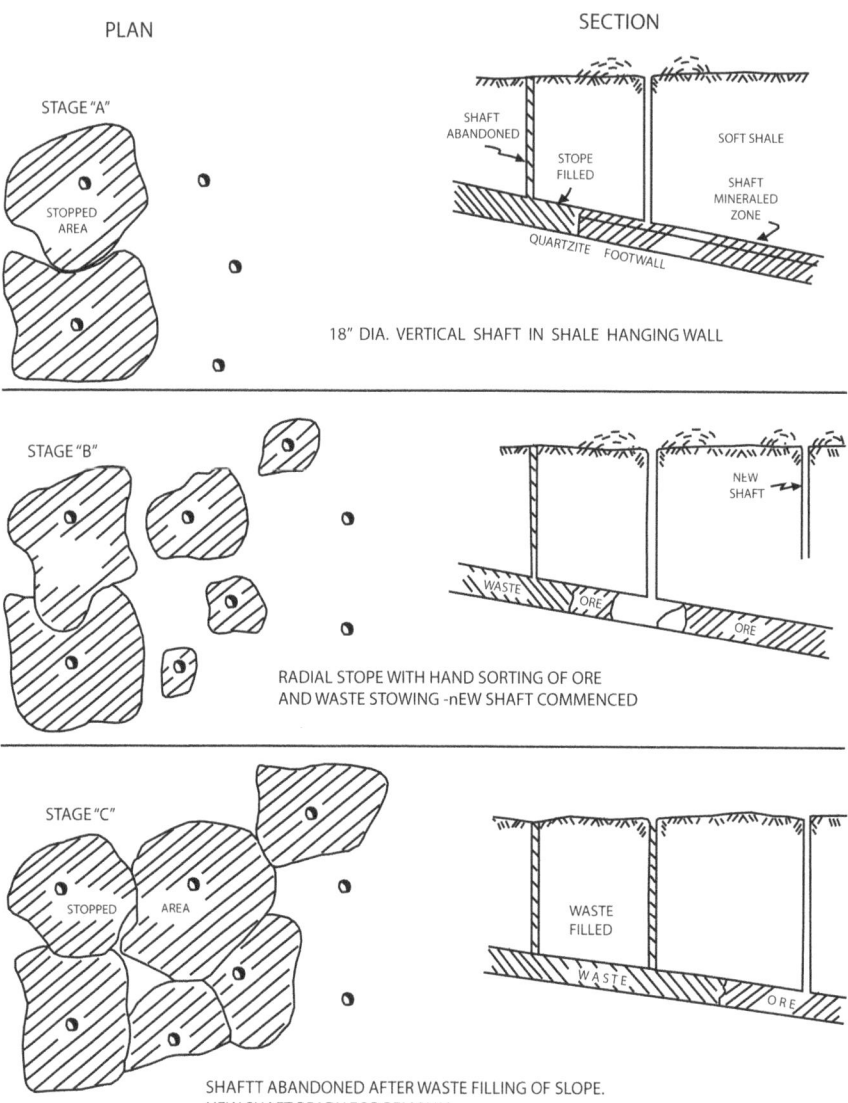

Fig. 3.4 Umkondo copper mine in Zimbabwe where miners strategically backfilled and opened up new shafts to create pillars (unmined blocks) for structural stability. (Redrawn from Summers 1969, p. 29, Fig. 6)

facilitate the use of one shaft as an up-draught chimney which drew air down the other one using the principle of convection (Miller 2000). These techniques were also used by New Kingdom Egyptians used for mining copper at Timna in Israel (Rothenberg 1962), showing that geological constraints homogenized approaches to ore extraction in many areas.

The Rooiberg mines also provide additional insights into the methods used to transport personnel and the ore to the surface (Recknagel 1906; Baumann 1919). In one of the ancient stopes, early mining geologists discovered an expertly cut tree whose branches had been strategically removed for ease of stepping during use as a ladder (Chirikure et al. 2007). Furthermore, some of the vertical shafts connecting to underground galleries and chambers had steps incised from the top to bottom for easy access to the work place.

The now destroyed precolonial copper mine on Lolwe Hill, Phalaborwa, represents one of the earliest dated underground mining complexes in sub-Saharan Africa. Van der Merwe and Scully (1971) dated charcoal from one of the galleries to the eighth century AD. Other absolute dates, material culture and historical evidence from in and around the mine suggest that the carbonate copper deposits on Lolwe Hill were exploited by local people up to the late nineteenth century (Van der Merwe and Scully (1971). As a site of modern copper mining operated by the Palabora Mining Company, the site is now one of the largest open cast mines ever developed in the world. The archaeology around Lolwe Hill suggests that at different points in time, beginning in the late first millennium AD until the nineteenth century, various communities dipped into the mines, but occupation was concentrated away from the source at places such as Kgopolwe (Van der Merwe and Scully 1971) and Shankare (Thondhlana 2012). It is possible that Early Iron Age copper mining on Lolwe Hill initially began with surface collection developed through open mining and ended up being a sophisticated underground operation.

During the archaeological research in the area, it was clear that the underground mines had started as open mines that only intersected rich veins and pockets of mineralization beneath the ground at depths of between 15 and 20 m (Miller et al. 2001). As the mineralization was mined out, a maze of galleries and adits was left underground. In one case, a wide vertical shaft ended abruptly in a round chamber, while in others inclined shafts branched off into horizontal galleries and adits (Van der Merwe pers comm 2013). Remarkably, Van der Merwe and his team observed that the miners at Lolwe followed the ore veins with great accuracy, resulting in the creation of narrow shafts underground which could only be worked by gracile people and or children. As demonstrated at Rooiberg and other places, child labour was probably widely used in African mining.

The mining engineering at Lolwe was very advanced, for the site was littered with very deep vertical shafts of not more than 45 cm in diameter that connected to underground galleries. These ventilation shafts provided air underground which was crucial for breathing and for driving away poisonous gases from fire setting and other mining activities. Inside the mine shafts and galleries, archaeologists found characteristic mining tools such as iron gads, chisels and dolerite hammerstones. A dense charcoal concentration found on the floor of yet another gallery on Lolwe Hill was presumably the result of fire setting which was often utilized to break hard rock from the gallery walls. The details of these precolonial workings on Lolwe Hill are not just evidence of the mining techniques used by the indigenous miners in the first and second millennium AD, but are also an indication of their geologically constrained and technologically inspired choices. Again, the underground mines in the

Egyptian desert (c. 4000 onwards) represent Africa's earliest underground mines which may have been worked by slaves (Klemm and Klemm 2012). As in later cases of mining, fire setting was an important technique for breaking the host rock.

The descriptions of underground mining in Africa have deep resonance and correspondence with the practices documented at other famous mining landscapes such as the copper mines at Timna in the Arabah desert of Israel (Hauptman 2007). Here, underground mining followed shafts and adits with good accuracy, while narrow ventilation shafts supplied air to the mining chambers underground and were important for sucking from underground the abundant poisonous gases. This picture was also observed at other early mining landscapes in places such as Cornwall in England during the Middle Ages. This similarity does not and should not necessarily imply the diffusion of ideas from multiple areas into Africa. Rather, it shows that when confronted with similar geological opportunities and constraints, ancient miners at different points in time, whether in Katanga in Democratic Republic of Congo, Zawar in India, Potosi in Bolivia, Cornwall in England and Sinai Desert in the Middle East selected technological choices that left identical results. Although there are many technological solutions (Sillar and Tite 2000), it seems that as far as mining is concerned and as far as geology dictates, there is one way in which below the ground ore can be mined regardless of time and place. Humanity had to sink shafts into the earth, and to make provisions for necessities such as air, and safety.

The problems affecting underground mining were mostly related to structural collapse (Fig. 3.3), flooding and lighting. The burning of logs and grasses may have provided lighting underground. Timbering inside the mines (e.g., at the Harmony Block), digging ventilation shafts (Lolwe and Timna) and creating pillars of unmined earth (Fig. 3.4) were technological and practical choices aimed at alleviating these problems. Arguably, the level of skill, technology and decision-making involved in underground mining is in excess of that invested in either alluvial or simple open mining.

With mining—whether surface, open or underground—the main challenge for archaeologists lie in establishing tight sequences of deposit exploitation. This is because later activities are always overwritten on earlier activities, such that in most cases the available dates are for more recent activities and not the earliest ones. Lolwe Hill is one of the few places where an adit was left with charcoal and other material characteristic of a single time horizon. It is also difficult to speculate what the earliest technique of mining was on the basis of perceived simplicity or complexity. This is because often all methods were simultaneously used by one group at the same locality (Hammel et al. 2000; Rothenberg 1999). As such, depending on needs, and other technological or cultural considerations, mining was a vast enterprise aimed at obtaining ores from the ground through surface collection, and digging underground.

Fig. 3.5 Wooden bucket found inside ancient gold mine in Southwestern Zimbabwe and donated to Natural History Museum, Zimbabwe. Exact provenance unknown. (Photo credit: Aluthor)

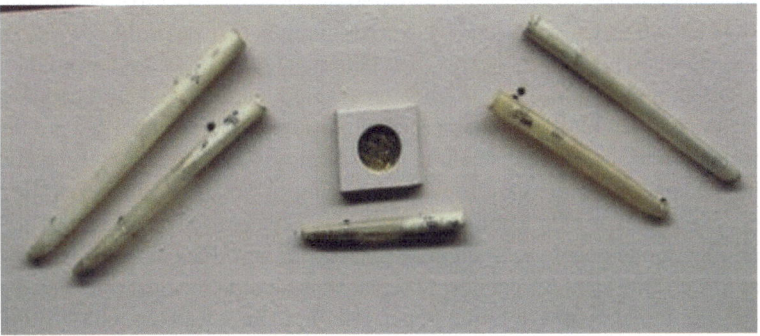

Fig. 3.6 Porcupine quills used for storing gold. (Photo Credit Author)

Preindustrial Hoisting and Beneficiation

In terms of haulage and hoisting, small wooden bowls and buckets (Fig. 3.5) were used to carry the ore and earth. Sometimes, buckets made of wood were attached to a rope and hoisted into deeper mines to transport the ore and gangue materials. In the case of Shona open gold mining, miners made up a human conveyor belt, passing the baskets from person to person (Ellert 1992). Sometimes, animal draught was used by the Njanja of Zimbabwe to carry sacks of ore from mines to smelting areas (Chirikure 2006). After processing, gold dust was placed in porcupine quills (Fig. 3.6) for later consolidation in crucibles. However, in times of increased

Fig. 3.7 Undated grinding stone with dolly holes used to crush copper ores at Phalaborwa. (Photo credit: Author)

demand, the Islamic traders operating at places such as Tewdaghoust often bought gold dust in quills (Curtin 1973), just as the Portuguese did in Southern Zambezia from AD1500 onwards.

One of the most important stages in the *chaîne opératoire* of mining was the process of beneficiating the ore, which involved cleaning it to remove occluded waste materials known as gangue. Once recovered, the miners dressed the ore to remove waste materials. This process is known as beneficiation, and it varied from metal to metal. Often, high-grade and low-grade ores were blended together to create more balanced ore in terms of exploitable metal and impurities essential for promoting slag formation in self-fluxing smelting technologies (Chirikure and Rehren 2004).

The evidence from the hard rock gold mining in Southern Zambezia suggests that fairly complex grinding and crushing methods were used to recover gold from its ore. The Shona people crushed gold-rich ore on hard rock surfaces where they ground it into fine sand for panning. The rock surfaces known as *ruware* in Shona acted as the female part while the upper grinding stone was the male (Summers 1969). This process left smoothly abraded but shallow depressions known as dolly holes (Fig. 3.7), some of which contained smears and traces of gold (Summers 1969; Swan 1994). In other cases, the ore was ground on grinding stones similar to the ones used for processing grain. The fine sand was then panned to recover gold dust which was then packed in porcupine quills (Fig. 3.6). According to Curtin (1973), such a technological solution was also practised at the Bambuk goldfields on the upper Senegal River in the first and second millennium AD.

At Rooiberg, in South Africa, the tin-rich host rock was crushed into small pieces which were ground on conventional grinding stones similar to those used for grain. The product from this process was then washed and panned to recover the cassiterite through density separation. Examples of such grinding stones were found in fairly large quantities at the large-scale tin-smelting site known as Smelterskop (Chirikure et al. 2010). Much larger rock-grinding stones were used in Zimbabwe gold mines and were illustrated by Summers (1969).

The literature on copper is not very detailed, but at Phalaborawa and Musina, where malachite is associated with magnetite, a very meticulous process of separating the iron from the copper carbonate was practised. Failure to achieve this resulted in co-smelting of iron and copper resulting in a very weak and useless copper iron alloy called *musina* by the Venda (Stayt 1931; see also Craddock and Meeks 1987 for a more technical description of iron in copper). The recovered material was ready for metallurgical processes such as smelting in clay-walled furnaces.

Klemm and Klemm (2012) published a number of mortars used for beneficiating gold ores in the Eastern Desert and Nubia. These mortars resemble the ones that were used in other parts of Africa, at much later periods suggesting that identical technological solutions were often applied in different contexts as dictated by the task at hand.

Mining Equipment and Other Paraphernalia

Generally, the tools and equipment used in preindustrial African mining served multiple purposes: excavating the earth, hoisting and haulage. Digging was accomplished using iron hoes attached to wooden handles, iron gads, chisels and wedges. Hammerstones were used to drive the chisels into the ore body and to break the resultant rocks into small-sized lumps. Shovels were important in scooping earth into carriers. From the Eastern Desert in Egypt to Nubia and Bambuk, through to Katanga and Phalaborwa in different parts of Africa, mining tools have been recovered by archaeologists, although preservation favoured inorganic equipment (Table. 3.1).

The Anthropology of Mining: A Global Outlook

In the late nineteenth and early twentieth centuries, there was an established belief in the Western academy that African technologies such as mining were shrouded in magic (Beuster 1881 cited in Rickard 1939; see Collett 1993 for critique). This contrasted significantly with perceptions of science-based Western technologies (e.g., Austen and Headrick1983). In this section, efforts are made to recap the cultural or, rather broadly speaking, the anthropological aspects of African mining and to compare them to practices elsewhere. The result demonstrates that, regardless of location, mining was a heavily ritualized process, requiring the intervention of deities, spirits of the land and the supernatural (Nash 1993).

When discussing cultural attributes associated with African mining, it is important to separate iron and copper from gold, and it is crucial to keep in mind that most graphic examples are from ethnographic cases, and the detailed recordings in Ancient Egypt and Nubia. The archaeological evidence is simply not that well preserved to allow a detailed reconstruction. The mining of copper and iron was variably associated with rituals and taboos. For example, amongst the Dogon of Mali,

Table 3.1 Techniques and tools used in precolonial mining across Africa. Adapted and modified after Hammel et al. (2002, p. 51)

Technique/Tool Description	Description	Purpose
Stone hammers and pounders	Dolerite, diorite or—more rarely—quartz-porphyry	Rock-breaker. Also used to 'greenstone' or granite drive gads
Iron gads/chisels	Smaller ones were pointed at both ends, one end being held in a wooden handle. Larger ones were driven with stone hammers	Used to split ore
Stone wedges	Very few examples from ancient mines. Wooden wedges, although unknown in the archaeological record due to their impermanence, would seem a more likely tool	Used to enlarge or deepen crack in rock
Fire setting	The process of heating rock and then cooling it rapidly (with water)	Used to crack rock
Digging Tools		
Hoe	A heavy diamond-shaped iron tool hafted in a wooden handle. Hoes worn down in the fields could be adapted for mine use	Used to break, dig or draw soil. Gathered together broken rock to draw into a basket, or other carrier
Shovel	Iron, with cranked shanks. Smaller iron scoops, known from Zambian, copper mines	To gather ore
Haulage and Hoisting		
Small wooden bowls	Not many examples. It seems that in shallow diggings, mined ore was passed in bowls from hand to hand until it reached the surface	Haulage in shallow mines
Buckets (wood/bark/leather/hide)	Fes examples, a small bucket (approx. 3-litre capacity) with three holes for suspension has been found (Hammel et al. 2000)	Haulage and hoisting in deeper mines
Baskets	Musina: evidence of baskets made of tightly woven palm ribs and reinforced with a leather cover	Carrying ore
Transport		
Baskets, sacks	Baskets made of vegetal material or sacks of leather	Carrying ore from mines to smelting places
Oxen	Sacks were placed on the back of oxen	Carrying ore from mines to smelting places
Sledges	Sacks and baskets loaded on a sledge for transportation	Carrying ore
Storage		
Porcupine quills	Quills used to store gold dust	Storage of gold dust for melting and trade
Pots	Clay pots	Carry ore, food drink for miners
Lighting		
Logs	Carbonized logs were found in adits suggesting their possible use as torches	Providing light

menstruating women were forbidden from the mines (Huysecom and Augustoni 1997). The presence of women in general too was limited or encouraged depending on context. Similarly, amongst the Phoka of Malawi, women were not allowed near mines and neither were they permitted to touch the ore (Van der Merwe and Avery 1987). This practice was also noted amongst various other groups such as the Toro in Uganda (Childs 1998). However, at Katanga in DRC, women of all ages participated in copper mining (Bisson 2000). Therefore, generalizations should not subsume context-specific variations.

Because mining involved crossing the boundary between the surface and the belly of the earth, it was seen as a dangerous activity, characterized by many unknowns. Miners across Africa and elsewhere often required the intervention of the deities via intermediaries such as spirit mediums. Medicines were an important part of the mining ritual, for good medicines kept malevolent influences at bay while making it easier to find ore. In the African ethnographic record, miners did not always manage to find good ores; there were instances of failure. Under such circumstances, offerings and sacrifices were made to the ancestors, e.g., in the form of beer or chicken. Once ancestors were propitiated, the earth would release the ore. Therefore, mining required the intervention of both the dead and the living to help cope with uncertainty. The involvement of ancestors was important for another reason; mining took place in nature, conceptually away from the realm of culture. In this liminal space, it was only the ancestors who could guide the living. Furthermore, amongst many societies, the land on which mines were located was also under the realm of ancestors, who made it rain, and made land fertile and productive.

Gold mining seems to have differed from copper and iron as far as rituals and taboos were concerned. It relied heavily on the labour of women and children and yet involved substantial underground excavation (Summers 1969). Why did different communities treat this metal differently? Presumably, this stemmed from the fact that gold exists in pure form in nature, thereby contrasting with copper and iron that underwent dangerous heat-mediated chemo-thermal transformations to produce usable metal. Often, as mentioned above, there were also some contexts where women worked inside iron and copper mines. Men and women were therefore different actors, who networked with each other within context given norms to achieve success. These rich contrasts warn researchers against making broad generalizations at the expense of culture-specific variation and milieus.

Arguably, precolonial African mining was simultaneously technological in as much as it was transformative and sociocultural. A brief comparison with well-documented cases in other parts of the world indicates an overwhelming presence of gods and supernatural forces in mining. Rothenberg and Bachmann (1988) discuss the Egyptian temple at Timna which catered for the needs of the miners, protecting them from harm. As hinted above, El Tio still plays and played an important role in ancient silver mining in South America (Nash 1993). The same applies to Nepal where religion and beliefs were central in technologies such as mining (Haaland 2004a). Therefore, magic was a strong feature of mining be it in Africa, Latin America and other parts of the world. It is only that most of the evidence has been destroyed while not enough recourse was made to historical sources to reclaim these important details (Schmidt 1997). Mining was embedded in the broader sociocultural fabric, and as such, it exhibited beliefs that were common in society: the belief in ancestors and deities; the power of evil forces; and concepts of pollution. Taking away this ritual from technological studies removes the oil that lubricated the system by empowering people to go beyond the limitations of the everyday so as to extract material from the other world. Today, we see ourselves as guided by technological concepts, but in the past, rituals and taboos played a fundamental role in routinizing learned behaviour in various processes, technical or otherwise.

Another important variable in preindustrial mining relates to the organization of labour. In preliterate contexts, it is difficult to fully understand how mining was organized without speculation. However, the Romans, and possibly the Egyptians, used slave labour in the gold mines of the Eastern Desert (Klemm and Klemm 2012). De Barros (1986) has argued that Bassar iron smelting was highly specialized with villages specializing in getting charcoal and ore, in smelting and in forging.

Conclusion

In conclusion, as the first stage in the chaîne opératoire of metal production and use, mining occupies an important place in technological and anthropological studies. The many rich metallogenic provinces of Africa contain substantive footprints of ore extraction spanning close to 5000 years old in Egypt and Nubia and 2000 years old in sub-Saharan Africa. By comparison, Timna, Faynan and other Eurasian landscapes have more layers but the similarity in techniques from surface collection through open to underground mining suggest that the geology to a large extent determined the techniques of mining and the technocultural solutions selected to cross the boundary between culture and nature. Far from promoting an unnecessary geological determinism, this view is supported by the observation that similar mineralization was worked using the same methods independent of spatial and temporal frameworks. For instance, if underground veins were excavated, there were high chances that the ensuing structural instability would result in collapse. This motivated various techniques of provisioning structural support in the mines, from the use of pillars, strategic backfilling to using wooden props. This can be seen from Phalaborwa, Katanga, in Africa to Potosi in the New World and Timna and Rio Tinto in Eurasia.

Finally, wherever humanity was mining ore in the world, rituals and belief systems formed part of the mining enterprise, and it is therefore myopic to think that this was a quintessentially African experience. The gods and spirits empowered the living to cross into a different conceptual space (underground). Therefore, pollution through menstruation or malevolent influences was unwelcome, just as these forces were feared in day-to-day life. There were risks associated with crossing this nature–culture boundary: death being one of the misfortunes that could befall miners. Because risk is both danger and opportunity, when miners managed to obtain the ore, how did they transform this commodity from the 'other world' into a cultural product for everyday use? Because the ore existed as a compound of different oxides, it was important to win elemental metal from the ore. This motivated for smelting to break the bond between the metal and the oxygen. This geochemical engineering was achieved through smelting in a heat and atmosphere regulated environment and is an important phase in the *chaîne opératoire* of metalworking, which is the focus of the next chapter.

References

Austen, R. A., & Headrick, D. (1983). The role of technology in the African past. *African Studies Review, 26,* 155–161.

Bandama, F. (2013). *The archaeology and technology of metal production in the Late Iron Age of the Southern Waterberg, Limpopo Province, South Africa*. Unpublished Doctoral dissertation, University of Cape Town.

Baumann, M. (1919). Ancient tin mines of the Transvaal. *Journal of the Chemical, Metallurgical and Mining Society of South Africa, 19,* 120–132.

Bellamy, C. V., & Harbord, F. W. (1904). A West African smelting house. *Journal of the Iron and Steel Institute, 66,* 99–126.

Bisson, M. (2000). Precolonial copper metallurgy: sociopolitical context. In M. Bisson, P. de Barros, T. C. Childs, & A. F. C. Holl (Eds.), *Ancient African metallurgy: The sociocultural context* (pp. 83–146). Walnut Creek: AltaMira Press.

Brown, J. (1995). *Traditional metalworking in Kenya*. Oxford: Oxbow.

Childs, S. T. (1998). 'Find the *ekijunjumira*': Iron mine discovery, ownership and power among the Toro of Uganda. In A. B. Knapp, V. C. Piggott, & E. W. Herbert (Eds.), *Social approaches to an industrial past: The archaeology and anthropology of mining* (pp. 123–137). Abingdon: Routledge.

Chirikure, S. (2006). New light on Njanja iron working: Towards a systematic encounter between ethnohistory and archaeometallurgy. *South African Archaeological Bulletin, 61,* 142–161.

Chirikure, S. (2007). Metals in society: Iron production and its position in Iron age communities of Southern Africa. *Journal of Social Archaeology, 7*(1), 72–100.

Chirikure, S. (2010a). *Indigenous mining and metallurgy in Africa*. Cambridge: Cambridge University Press.

Chirikure, S., & Rehren, T. (2004). Ores, furnaces, slags, and prehistoric societies: Aspects of iron working in the Nyanga agricultural complex, AD 1300–1900. *African Archaeological Review, 21*(3), 135–152.

Chirikure, S., & Rehren, T. (2006). Iron smelting in pre-colonial Zimbabwe: Evidence for diachronic change from Swart village and Baranda, Northern Zimbabwe. *Journal of African Archaeology, 4,* 37–54.

Chirikure, S., Hall, S., & Miller, D. (2007). One hundred years on: what do we know about tin and bronze production in Southern Africa? In: S. La Niece, D. Hook, P. Craddock (Eds.), *Metals and mines, studies in archaeometallurgy* (pp. 112–119). London: British Museum Press.

Chirikure, S., Heimann, R. B., & Killick, D. (2010). The technology of tin smelting in the Rooiberg Valley, Limpopo Province, South Africa, ca. 1650–1850 CE. *Journal of Archaeological Science, 37*(7), 1656–1669.

Cline, W. B. (1937). *Mining and metallurgy in Negro Africa* (No. 5). Menasha: George Banta.

Collett, D. P. (1993). Metaphors and representations associated with precolonial iron—smelting in Eastern and Southern Africa. In T. Shaw, P. Sinclair, B. Andah, & A. Okpoko (Eds.), *The archaeology of Africa: Food, metals and towns* (pp. 499–511). London: Routledge.

Craddock, P. T. (1995). *Early metal mining and production*. Washington, DC: Smithsonian University Press.

Craddock, P. T. (2000). From hearth to furnace: Evidences for the earliest metal smelting technologies in the Eastern Mediterranean. *Paléorient, 26*(2), 151–165.

Craddock, P. T., & Meeks, N. D. (1987). Iron in ancient copper. *Archaeometry, 29*(2), 187–204.

Curtin, P. D. (1973). The lure of Bambuk gold. *Journal of African History, 14*(4), 623–631.

Dart, R. A., & Beaumont, P. (1967). Amazing antiquity of mining in Southern Africa. *Nature, 216*, 407–408.

David, N., Heimann, R., Killick, D., & Wayman, M. (1989). Between bloomery and blast furnace: Mafa iron-smelting technology in North Cameroon. *African Archaeological Review, 7*(1), 183–208.

de Barros, P. (1986). Bassar: A quantified, chronologically controlled, regional approach to a traditional iron production centre in West Africa. *Africa, 56*(02), 152–174.

de Barros, P. (1988). Societal repercussions of the rise of large-scale traditional iron production: A West African example. *African Archaeological Review, 6*(1), 91–113.

de Barros, P. (2013) A comparison of early and later Iron Age societies in the Bassar region of Togo. In J. Humpris & Th. Rehren (Eds.), *The world of iron* (pp. 10–21). London: Archetype.

Ellert, H. (1992). *Rivers of gold*. Gweru: Mambo Press.

Evers, T.M., & Van den Berg, R.P. (1974). Ancient mining in Southern Africa, with reference to a copper mine in the Harmony Block, North-Eastern Transvaal. *Journal of the South African Institute of Mining and Metallurgy, 74*, 217–226.

Friede, H. M., & Steel, R. H. (1976). Tin mining and smelting in the Transvaal during the Iron Age. *Journal of the South African Institute of Mining and Metallurgy, 76*(12), 461–470.

Garrard, T. F. (2011) (1989). *African Gold: Jewellery and ornaments from Ghana, Côte d'Ivoire, Mali and Senegal in the collection of the Barbier-Mueller Museum*. Munich: Prestel.

Haaland, R. (2004a). Technology, transformation and symbolism: Ethnographic perspectives on European iron working. *Norwegian Archaeological Review, 37*(1), 1–19.

Haaland, R. (2004b). Iron smelting—a vanishing tradition: Ethnographic study of this craft in South-West Ethiopia. *Journal of African Archaeology, 2*(1), 65–79.

Hall, S. L. (1981). *Iron Age sequence and settlement in the Rooiberg, Thabazimbi area*. Masters thesis, University of the Witwatersrand, Johannesburg.

Hammel, A., White, C., Pfeiffer, S., & Miller, D. (2000). Pre-colonial mining in Southern Africa. *Journal of South African Institute of Mining and Metallurgy, 100*(1), 49–56.

Hauptmann, A., Heitkemper, E., Pernicka, E., Schmitt-Strecker, S., & Begemann, F. (1992). Early copper produced at Feinan, Wadi Araba, Jordan: The composition of ores and copper. *Archeomaterials, 6*(1), 1–33.

Herbert, E. W. (1993). *Iron, gender, and power: Rituals of transformation in African societies*. Bloomington: Indiana University Press.

Holl, A. F. (2009). Early West African metallurgies: New data and old orthodoxy. *Journal of World Prehistory, 22*(4), 415–438.

Huffman, T. N., Van der Merwe, H. D., Grant, M. R., & Kruger, G. S. (1995). Early copper mining at Thakadu, Botswana. *Journal of the South African Institute of Mining and Metallurgy, 95*, 53–61.

Huysecom E., & Agustoni B. (1997). *Inagina: l'ultime maison du fer/the last house of iron*. Geneva: Telev. Suisse Romande. (Videocassette, 54 min).

Ige, A., & Rehren, Th. (2003). Black sand and iron stone: Iron smelting in Modakeke, Ife, South-Western Nigeria. *Institute for Archaeo-Metallurgical Studies, 23*, 15–20.

Killick, D. (2004b). Social constructionist approaches to the study of technology. *World Archaeology, 36*(4), 571–578.

Killick, D., & Miller, D. (2014). Smelting of magnetite and magnetite-ilmenite iron ores in the Northern Lowveld, South Africa, ca. 1000 CE to ca. 1880 CE. *Journal of Archaeological Science, 43*, 239–255.

References

Killick, D., Van der Merwe, N. J., Gordon, R. B., & Grébénart, D. (1988). Reassessment of the evidence for early metallurgy in Niger, West Africa. *Journal of Archaeological Science, 15*(4),367–394.

Klemm, R., & Klemm, D. D. (2012). *Gold and gold mining in ancient Egypt and Nubia: Geoarchaeology of the ancient gold mining sites in the Egyptian and Sudanese eastern deserts*. New York: Springer.

Levtzion, N. (1973). *Ancient Ghana and Mali* (Vol. 7). London: Methuen.

Mackenzie, J. M. (1975). A pre-colonial industry: The Njanja and the iron trade. *Nada, 11*(2), 200–220.

Mamadi, M. F. (1940). The copper miners of Musina. In N. J. van Wamelo (Ed.), *The copper miners of Musina and the early history of the Zoutpansberg. Ethnological Publications VIII* (pp. 81–87). Pretoria: Department of Native Affairs, Union of South Africa.

Mbiti, J. S. (1990). *African religions & philosophy*. Portsmout: Heinemann.

Miller, D. (2000). 2000 years of indigenous mining and metallurgy in Southern Africa—a review. *South African Journal of Geology, 98,* 232–238.

Miller, D. E., & Hall, S. L. (2008). Rooiberg revisited–the analysis of tin and copper smelting debris. *Historical Metallurgy, 42*(1), 23–38.

Miller, D., & Killick, D. (2004). Slag identification at Southern African archaeological sites. *Journal of African Archaeology, 2*(1), 23–47.

Miller, D., Killick, D., & Van der Merwe, N. J. (2001). Metal working in the Northern Lowveld, South Africa AD 1000–1890. *Journal of Field Archaeology, 28*(3–4), 3–4.

Nash, J. C. (1993). *We eat the mines and the mines eat us: Dependency and exploitation in Boliviantin mines*. New York: Columbia University Press.

Phimister, I. R. (1974). Alluvial gold mining and trade in nineteenth-century south central Africa. *The Journal of African History, 15* (3), 445–456.

Pikirayi, I. (2001). *The Zimbabwe culture: Origins and decline in Southern Zambezian states*. Vol. Walnut Creek: Altamira.

Radivojević, M., Rehren, T., Pernicka, E., Šljivar, D., Brauns, M., & Borić, D. (2010). On the origins of extractive metallurgy: New evidence from Europe. *Journal of Archaeological Science, 37*(11), 2775–2787.

Recknagel, R. (1906). On some mineral deposits in the Rooiberg district. *Transactions of the Geological Society of South Africa, 11,* 83–106.

Rickard, T. A. (1939). The primitive smelting of iron. *American Journal of Archaeology, 43*(1), 85–101.

Robb, L. (2009). *Introduction to ore-forming processes*. London: Wiley.

Robion-Brunner, C, Serneels, V, & Perret, S. (2013). Variability in iron smelting practices: assessment of technical, cultural and economic criteria to explain the metallurgical diversity in the Dogon area (Mali). In J. Humpris & Th. Rehren (Eds.), *The world of iron* (pp. 257–265). London: Archetype.

Rothenberg, B. (1962). Ancient copper industries in the Western Arabah: An archaeological survey of the Arabah, Part 1. *Palestine Exploration Quarterly, 94*(1), 5–71.

Rothenberg, B. (1999). Archaeo-metallurgical researches in the Southern Arabah 1959-1990. Part 2: Egyptian new kingdom (Ramesside) to early Islam. *Palestine exploration quarterly, 131*(JULDEC), 149–175.

Rothenberg, B., & Bachmann, H. G. (1988). *The Egyptian mining temple at Timna*, (Vol. 1). London: Institute for Archaeometallurgical Studies.

Schmidt, P. R. (1997). *Iron technology in East Africa: Symbolism, science, and archaeology*. Bloomington: Indiana University Press.

Schmidt, P. R. (2009). Tropes, materiality, and ritual embodiment of African iron smelting furnaces as human figures. *Journal of Archaeological Method and Theory, 16*(3), 262–282.

Sillar, B., & Tite, M S. 2000. The challenge of 'technological choices' for materials science approaches in archaeology. *Archaeometry, 42,* 2–20.

Stayt, H. A. (1931). *The Bavenda* (No. 58). International Institute of African languages & cultures. Oxford: Oxford University Press.

Summers, R. (1969). *Ancient mining in Rhodesia and adjacent areas*. Salisbury: Trustees of the National Museums of Rhodesia.

Swan, L. (1994). *Early gold mining on the Zimbabwean plateau*. Studies in African Archaeology, 9. Uppsala: Societas Archaeologica Upsaliensis.

Thondhlana, T. (2012). *Metalworkers and smelting precincts: Technological reconstructions of second millennium copper production around Phalaborwa, Northern Lowveld of South Africa*. Unpublished Doctoral dissertation, University College of London.

Van der Merwe, N. J., & Avery, D. H. (1987). Science and magic in African technology: Traditional iron smelting in Malawi. *Africa, 57*(2), 143–172.

Van der Merwe, N. J., & Scully, R. T. (1971). The Phalaborwa story: Archaeological and ethnographic investigation of a South African Iron Age group. *World Archaeology, 3*(2), 178–196.

Wagner, A., & Gordon, S. (1929). Further notes on ancient bronze smelters in the Waterberg district, Transvaal. *South African Journal of Science, 26,* 563–574.

Watts, I. (2002). Ochre in the Middle Stone Age of Southern Africa: Ritualised display or hide preservative? *The South African Archaeological Bulletin, 57,* 1–14.

Willies, L., Craddock, P. T., Gurjar, L. J., & Hegde, K. T. M. (1984). Ancient lead and zinc mining in Rajasthan, India. *World Archaeology, 16* (2), 222-233.

Chapter 4
Domesticating Nature

Introduction: Transforming Ore into Metal

According to Herbert (1993), the role of ancestors, ritual, and the supernatural in the pre-industrial African metal production chain did not end with the process of mining. Instead, the power of ritual, at least in the ethnography of sub-Saharan Africa, was important in the reduction of oxide or carbonate ores to metal. The longevity of this practice is suggested by excavation of furnaces dating to the first and second millennium AD associated with holes where medicines were strategically planted at the bottom of the iron smelting furnaces in Central, Eastern and Southern Africa (Mapunda 1995; Rowlands and Warnier 1993; Schmidt 1997) (Fig. 4.1). In Egypt and Nubia, Gods such as Maat were important throughout the history of metalworking in the ancient past (Scheel 1989) and smelting was often done within temple precincts (El Rahman et al. 2013). Within the context of glass making, Robson (2001, p. 54) argued that the boundaries between science and religion, medicine and magic were always blurred in the ancient Near East: The spiritual was inseparable from the rational. This demonstrates continuity in various practices across Africa but does not in any way suggest that African extractive metallurgy was static through time.

Smelting is a fundamental step in the *chaîne opératoire* of metal production. For the ore to be successfully tamed into a cultural product—metal—great skill was needed in selecting and assembling suitable raw materials such as ore, air (supplied by blowing, bellows or naturally using the principle of convection), clay for making combustion vessels and receptacles used during smelting and charcoal fuel. These raw materials are generic and cross-cut metals worked pre-industrially. Food, labour and in some cases music were important but archaeologically invisible components of the metal production process (Dewey 1991).

Preindustrial extractive metallurgy techniques differed significantly from the industrial processes now currently in use (Rostoker and Bronson 1990). The production of iron in modern blast furnaces passes through two stages: (1) reduction of high-grade iron ores using coke and limestone flux to produce liquid cast iron and

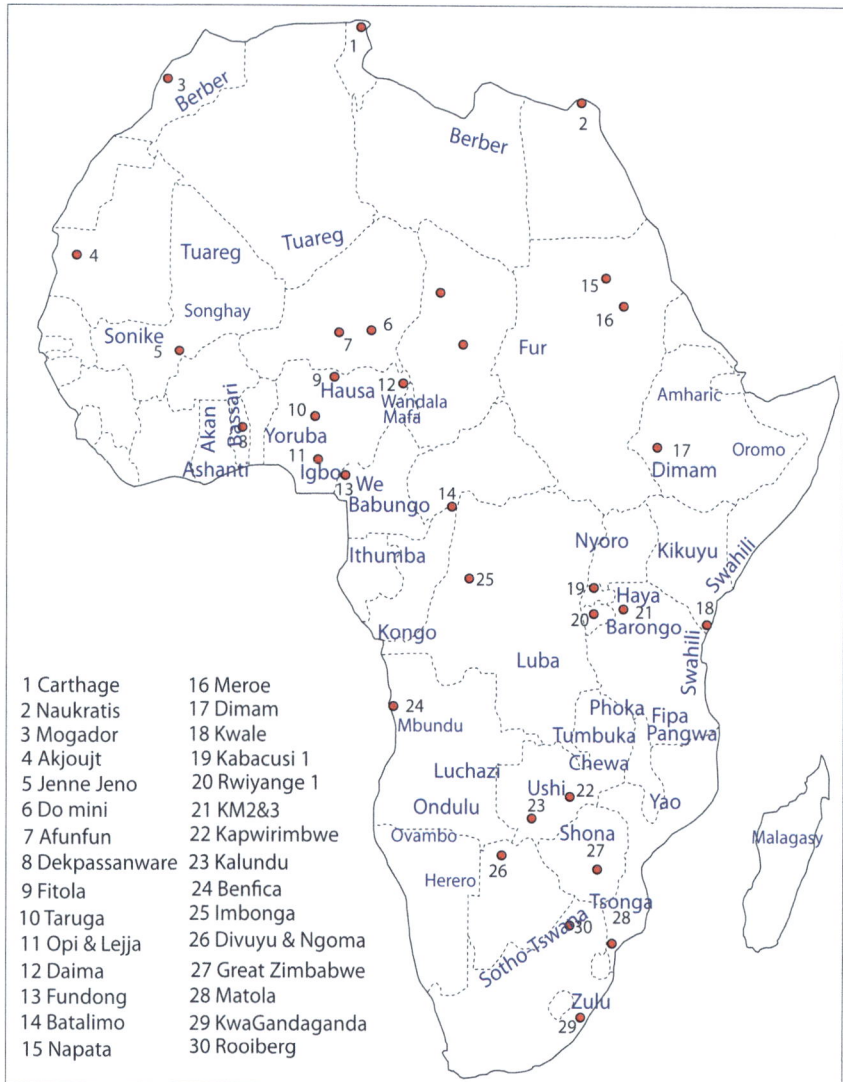

Fig. 4.1 Known metal smelting groups in Africa. *Numbers* indicate key sites and landscapes

very glassy slag and (2) further processing of cast iron in reveberatory furnaces to moderate carbon levels to create usable metal ranging from wrought iron to medium-and high-carbon steels (Rostoker and Bronson 1990). This indirect technique differs significantly from the bloomery process where ores were reduced to produce solid metal and liquid slag in a single operation (Tylecote 1980). This 'direct method', as it is known, was the principal method of metal reduction for millennia in much of the Old World, with the exception of China which had the blast furnace method from early on Wagner (2008).

The inventory of metals smelted in preindustrial Africa varied from region to region. In Egypt, Nubia and North Africa, the seven metals of antiquity were worked: gold, copper, silver, tin, lead, iron and mercury. Because of its different history, the metals worked in sub-Saharan Africa were limited to four: iron and copper (first millennium BC to recent times i.e. AD 1900) and, from the second millennium AD, tin and gold (Cline 1937; Killick 2014). Some sources suggest that lead was worked in the Benue region of Nigeria in the late first and early second millennium AD and in the Democratic Republic of Congo between the seventeenth and nineteenth centuries (Bisson 2000; Chikwendu et al. 1989). These metals were worked in diverse furnace types that differed from group to group, time to time and area to area. Often, copper and tin were mixed to produce an alloy known as bronze in Egypt since the Old Kingdom, as on the Jos plateau of Nigeria (c. AD 900–1200) and in Rooiberg in South Africa (c. AD 1450–1850). Because gold is a noble metal that exists in metallic state in nature, it was treated differently from iron, copper and tin. Instead, the gold dust or nuggets from the mines were melted in ceramic crucibles to consolidate them into usable pellets (Ellert 1993; Nixon et al. 2011; Ogden 2000; Oddy 1984; Summers 1969).

The waste products from smelting are remains of high-temperature processes, which contain within their chemical composition and microstructures partial histories of the processes that they have undergone (Bachmann 1982; Craddock 1995; Rehren et al. 2007). Not surprisingly, they form the staple of archaeometallurgy—a subdiscipline of archaeology that studies metal production and consumption in past societies (Rehren and Pernicka 2008; Roberts et al. 2009). Archaeometallurgical studies have primarily been concerned with reconstructing the technology of the process as revealed by slags, tuyeres, collapsed furnaces and remnants of ore (Bachmann 1982; Bayley and Rehren 2007; Hauptmann 2000; Heimann et al. 2010; Morton and Wingrove 1969). Such studies shed light on the quality of the ore used (Killick 1990), the efficiency of reduction (metal recovery) (Chirikure 2005, 2006) and the refractory nature of the clays used to make technical ceramics (tuyeres and furnaces) (Miller and Killick 2004). Because archaeometallurgical methods of investigation are rooted in earth and engineering sciences, until recently, the results of this technique were highly technical, largely ignoring the cultural attributes of the technology and its role in society (Rehren et al. 2007). Today, however, it is recognized that any study of preindustrial metalworking must focus on the materials and their materiality in order to understand the technology in its social context. The rest of the chapter is organized as (1) a brief outline of the raw materials needed for smelting, (2) a discussion of the chemistry of reduction, (3) a brief overview of metal smelting in West, Central, Southern, East and North Africa is provided within a diachronic framework, after which (4) the chapter concludes with a discussion of the anthropology of smelting.

Raw Materials

For reduction to succeed, and for smelters to transform nature into culture, they had to gather all the fundamental raw materials. These include the clay for fashioning infrastructure to hold the charge, bellows to pump air, and charcoal whose combustion was essential for initiating and sustaining endothermic reactions and reducing the ore to metal. Suitable ore was indispensable, for its reduction produced metal. The available evidence shows changing preferences in raw material selection as ancient metallurgists became more and more skilled in metal production (Craddock 2000).

Successful reductive smelting demanded ores of sufficient quality which, as we saw in Chap. 3, were sourced through various forms of mining. With respect to iron, a wide variety of ores were worked across Africa, from hematite at Dekpassanware in Togo during the Early Iron Age (c. 400 BC to AD 1000) (de Barros 2013), through ilmenite-rich magnetite sands in Yorubaland (c. AD 1700–1850) (Ige and Rehren 2003), to laterites in Nyanga Northeastern Zimbabwe (AD 1700–1850) (Chirikure and Rehren 2004) (Fig. 4.1). In terms of copper, malachite, cuprite and azurite were the most frequently used types across all ages from Wadi Dara in Egypt c. 4000 BC (Craddock 2000) to Musina in Venda, South Africa, in the nineteenth century. Sulfide ores were generally avoided in most of sub-Saharan Africa (Bisson 2000; Miller and Killick 2004), but were worked in Egypt and Nubia. Cassiterite predominantly appears to be the only tin ore that was smelted in sub-Saharan Africa and possibly Egypt (El Rahman et al. 2013; Heimann et al. 2010). Overall, the different ores that were smelted and the constraints and opportunities which they conferred to smelters often resulted in the development throughout the history of metallurgy in Africa of localized and context-specific smelting techniques and recipes (Chirikure and Bandama 2014).

Wood charcoal harvested and processed in a variety of ways acted as fuel whose combustion supplied heat to spur chemothermal reactions in the furnaces. Charcoal was produced through the incomplete incineration of wood (Horne 1982). Most of what we know about charcoal comes from the ethnographic and historical records because very little work has been done on the anthracology of smelting sites (Lyaya 2013; Mapunda 1995). In the twentieth century, the Dogon of Mali cut down dry trees and burned them before covering the inferno with sand to create a reducing environment (Huysecom and Augustoni 1997). This dry distillation of wood produced charcoal of good quality, and the method was also utilized by the Shona of Zimbabwe and the Kaonde of Zambia in the recent past (Cline 1937), but methods of producing charcoal in antiquity have not been explored in full. Wood sources had to be managed because overharvesting resulted in environmental and ecological degradation. The ratio of wood to charcoal in some West African communities was 10 to 1 (Goucher 1981), such that in the Mema region of ancient Ghana (modern day Mali), intensive iron smelting in the late first millennium AD severely depleted trees resulting in massive soil erosion (Haaland 1980), and there appears to have been a reduction in forest cover around Begho in modern Ghana (Goucher 1981). However, Njanja metal workers of Zimbabwe managed the woodlands through alternating tree species, thereby avoiding upsetting the ecological balance as was the case with Mema (Chirikure 2006). In Europe, intensive metal production was associated with negative ecological consequences, such as the depletion of forests that stimulated woodland management techniques such as coppicing (Joosten et al. 1998).

Smelting required access to good clays for making charge receptacles and tuyeres (blow pipes) for supplying air into the furnaces. The types of clays used varied from area to area, but in general, they had to be sufficiently refractory to maintain mechanical integrity of the furnace during smelting and at the same time gradually melt to contribute to slag formation (David et al. 1989; Martinón-Torres and Rehren 2014). Furthermore, selected clays were supposed to keep furnaces insulating enough to keep heat loss during reduction to a minimum (Crew 1991). It is possible that crucible furnaces were used in the early stages of copper smelting in Egypt and the Middle East (Craddock 2000; Hauptmann 2007). Smelting furnaces used in precolonial Africa have been grouped into four broad classes: crucible, bowl, low shaft, and high shaft types (Chirikure et al. 2009; Kense 1985; Miller and Van der Merwe 1994; Van der Merwe 1980; see Fig. 4.2 for distribution). Crucible furnaces comprised ceramic vessels that were fired from inside using blowpipes and were mostly used in ancient Egypt (Ogden 2000). In sub-Saharan Africa, crucible furnaces were used for smelting copper at Kansanshi in Zambia between AD 1000 and 1200 (Bisson 2000). Bowl furnaces consisted of a pit without any protruding superstructure, and low shaft and tall shaft furnaces stood to a height not exceeding 1, 5 and 7 m respectively (Chirikure et al. 2009). These are merely broad categories and contain variations within each group.

The archaeology of sub-Saharan Africa suggests a broad developmental trajectory from the non-slag tapping shaft and bowl furnaces used in the first millennium BC to the introduction of natural draught furnaces by the late first millennium AD (de Barros 2013; Robion-Brunner et al. 2013). The temperatures for reducing varied ores in these equally diverse furnaces ranged between 1100 and 1200°C. The furnaces that were worked when sub-Saharan African metallurgy began somewhere in the first millennium BC are easily identifiable by slag blocks and were non-slag tapping (Alpern 2005). Bowl furnaces consisted of a semicircular depression in the ground lined with refractory materials. A variant of this type had superimposed short shafts aimed at providing high volumes and better draught when compared to the ordinary bowl type (Miller and Van der Merwe 1994). Bowl furnaces were both slag tapping and non-slag tapping (Ackermann et al. 1999; Cline 1937). Slag tapping is the process of draining the molten slag out of the furnace as the smelting process unfolded and had the advantage that furnaces could be used continuously and that more output could be obtained without the furnace getting full (Craddock 1995). In contrast, non-slag tapping furnaces had no such provision. In some designs, the slag drifted down into a specially designed pit (slag pit furnaces) or was raked out after the smelting was complete. The low shaft furnace type stood to between one and one and half meters above the ground and the diameter at the base too varied (Kense 1985; see David et al. 1989 for a 2.7-m-high exception). The shaft acted as the combustion vessel and was insulating enough to promote heat retention during smelting. Further distinctions have been made on these low shaft furnaces between those that had a provision for slag tapping and those without this feature (Van der Merwe 1980). The third group mostly consisted of high shaft furnaces that stood between 1.5 and 4 m above the ground. In contrast to the bowl and low shaft varieties that were operated by forced draught, these huge furnaces were

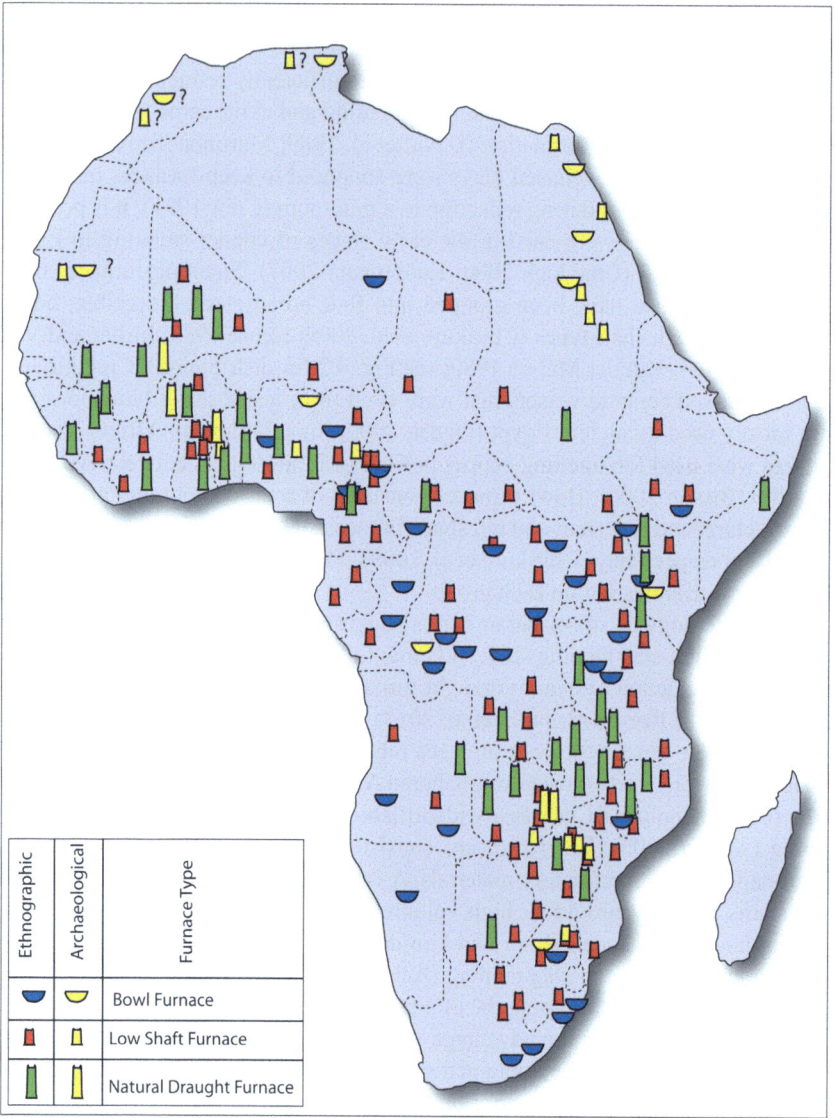

Fig. 4.2 Approximate distribution of bowl, shaft and natural draught furnace types in Africa. (From Chirikure et al. 2009)

universally powered by natural draught (Chirikure et al. 2009; Kense 1985; Van der Merwe 1980). Some natural draught furnaces used in West and Central Africa had a provision for slag tapping (e.g., Killick 1990, 1991), while others were non-slag tapping (Huysecom and Augustoni 1997). Archaeologically, it is difficult to distinguish between different types of shaft furnaces, but often natural draught furnaces have tuyeres fused in multiples (Prendergast 1975).

Slag tapping is seen as an advanced feature of smelting in Egypt from Ramesside times. In Europe, it was mostly popular from Roman times onwards (Pleiner 2000). Within an evolutionary perspective, it appears that slag removal in Central European furnaces passed through various stages. The first involved slag solidification inside the furnaces, while the second involved slag pits that collected slag underneath the furnaces. The third process of tapping appeared much later (Joosten et al. 1998; Pleiner 2000).

Copper and perhaps tin were smelted in bowl and low shaft furnaces, employing tapping and non-slag tapping technologies. These metals were rarely smelted in natural draught furnaces because they are highly reducing, which may have produced unwanted alloys of copper and iron and tin and iron (Craddock and Meeks 1987; Killick 1991).

The air for initiating and sustaining combustion was generated either artificially by blowing (in ancient Egypt) and pumping bellows (forced draft furnaces) or naturally by letting the furnace act as a chimney, which drew in air using the principle of convection (natural draught furnaces). In general, two types of bellows were used in precolonial Africa: bag and pot/drum varieties (see Fig. 4.3). A variant of the pot type known as concertina bellows was restricted to West Africa and was very efficient in generating air (Cline 1937). The bellows were connected to clay tuyeres or blow pipes, which directed air to the furnace. In Uganda, the Toro people gendered the tuyeres as male (Childs 1998). Schmidt and Avery (1983) believe that preheating the air in the tuyeres raised temperatures resulting in the production of high-carbon steels (see Rehder 1986 for alternative view). One of the earliest evidence of bellows used in smelting comes from Egypt and Egyptian paintings that show drum bellows from the third millennium BC onwards. At Meroe in the Sudan, Shinnie (1985) excavated an iron smelting furnace (Furnace 5) dating to AD 300 with pots used as bowls still in place. Elsewhere in Africa, bellows have rarely survived, particularly in cases where they were made using perishable materials.

In summary, the work carried out in different parts of Africa showed that sources of raw materials may have changed from time to time just as the furnace types and their method of operation were not static over time. Different scales of production motivated for the development of new innovations such that the story of African metallurgy is one of dynamism and change.

Brief Overviews: Metal Smelting in Preindustrial Africa— Egypt, Nubia, North Africa and the Horn of Africa

Chronology of Egypt and Nubia The earliest metal smelting evidence in Africa comes from Egypt. The very first Egyptian metal artifacts were recovered from Neolithic settlements such as Badari situated south of modern Asyut in Middle Egypt and were probably forged from native copper (Ogden 2000; Scheel 1989). Most of the objects recovered from Badarian sites were isolated objects from burial contexts. The evidence from Naqada, situated about 27 km (17 miles) north of

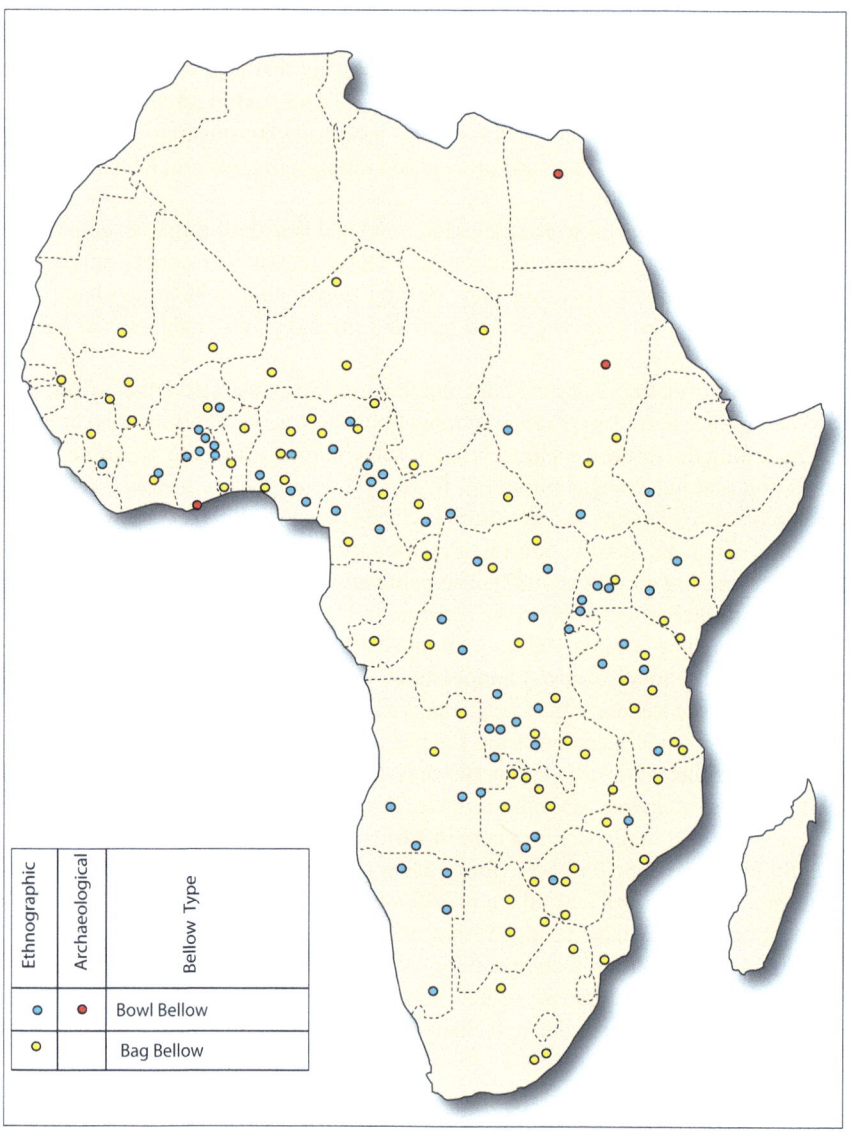

Fig. 4.3 Distribution of bellows across sub-Saharan Africa. (From Chirikure et al. 2009)

modern Luxor, shows that during the predynastic (Table 4.1), between c. 4000 and 3000 BC, metal processing became more common during the chalcolithic Naqada 1–111 phases (Emery 1970; Scheel 1989). In the Naqada 11 and 111 phases (about 3500 to 3050 BC), the evidence for copper smelting became more common. Naqada I1 culture spread over the entire Nile valley from north of Hierakonpolis into the Delta, and Egypt was networked with the surrounding regions via long-distance trade (Scheel 1989) (Figs. 4.4 and 4.5).

Table 4.1 Chronology of ancient Egypt and Nubia. (After Scheel 1989)

Period	Dates
Predynastic	5050–3050 BC
Old kingdom	2613–2181 BC
New kingdom	1570–1070 BC
Late period	713–332 BC
Ptolemaic period	332 BC–AD 395

Fig. 4.4 Location of early Egyptian and Nubian smelting sites. (After Scheel 1989)

The earliest stages in Egyptian extractive metallurgy are represented by a non-slagging process where reduction took place in crucibles. This was followed by a slagging process evident in the early third millennium BC at smelting sites such as Wadi Dara in the eastern desert of Egypt as well as somewhat further afield at Timna and Faynan (Craddock 2000). From the beginning of the dynastic period in Egypt at about 3050 BC, metal smelting techniques were continuously developed and refined. With the centralization of the Egyptian administration and the formation of a cultural centre at the royal capital, various professions and trades were established.

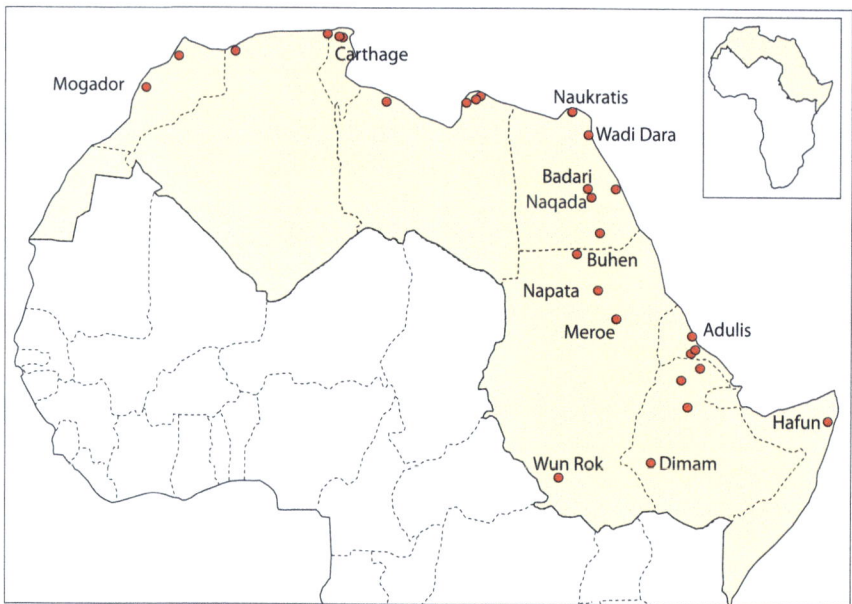

Fig. 4.5 Location of sites with smelting evidence in North Africa, Egypt, Nubia and Ethiopia

The first pictorial and inscriptional sources relating to the metalworker's craft come from Egyptian mastaba tombs at the beginning of the Old Kingdom (c. 2600 BC) (Scheel 1989). The first scenes of metalworking were found in Giza, but more pictorials appear on the tombs of officials in all periods of Egyptian history. Pictorial, inscriptional and archaeological sources, including the metal artifacts themselves, together serve as the basis for our investigation into Egyptian extractive metallurgy.

The furnaces for smelting copper evolved over time. During the Old and Middle Kingdom, smelting took place in crucibles appearing in the form of ceramic bowls (Fig. 4.6). Bowl furnaces developed over time. In the course of the smelting process, a mixture of crushed malachite and charcoal—the charge—was roasted and reduced to small prills of rich copper ore embedded in a copper slag. The prills were extracted by crushing the slag and then were melted together to form copper ingots. In Ramesside times, sophisticated pot bellow-driven shaft furnaces made of bricks were used, which enabled higher temperatures of about 1200°C to be reached (Fig. 4.7). These brick-built furnaces were fitted with a tap hole for draining slag (Scheel 1989). In the course of the smelting process, the copper droplets sank to the furnace bottom, forming an ingot. At the end of the process, the lighter slag above the smelted metal could be tapped into a slag pit through the tap hole in the furnace wall. Afterwards, the copper ingot was removed from the bottom of the furnace. The methods of air supply also changed with time, from the use of foliage in the predynastic period to the use of simple mouth blow piece made of reed and tipped with clay (Scheel 1989). During the Middle Kingdom, skin bellows of a goat or gazelle actuated by feet and cords were used with drum, pot and dish bellows.

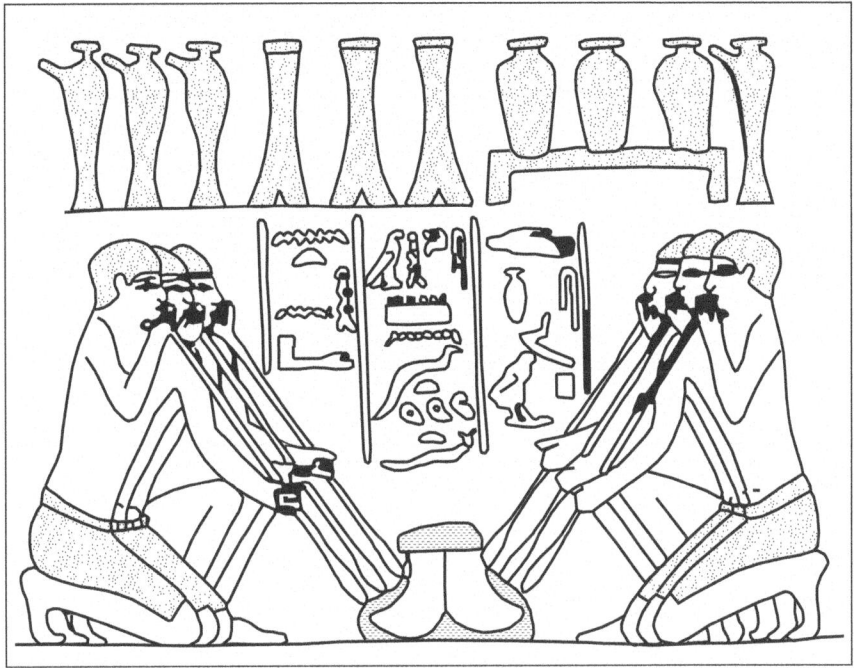

Fig. 4.6 An Old Kingdom pictorial showing six smelters blowing into two crucibles (Mastaba of Mereruka, fifth Dynasty). (Redrawn from Duel 1938, plate 30)

Although meteoric iron is known in Egypt from c. 3000 BC, the actual reductive smelting of ores to produce usable iron was very late. Unlike other Middle Eastern areas which adopted iron after 1500 BC, the earliest indications of iron smelting in Egypt were found in the delta region, particularly at the sites of Naucratis (Figs. 4.4, 4.5) and Defenna where archaeologists recovered a large quantity of iron slag and some ore. The site of Naucratis can be dated to about 580 BC. The iron was worked by Greek and Carian mercenaries after the expulsion of the Kushites (Arkell 1966; El Rahman et al. 2013). Iron was also smelted in the central eastern desert at Wadi Abu Gerida as shown by the presence of furnaces and olivine-rich slags dating to Ptolemaic period (El Rahman et al. 2013; Ogden 2000).

In lower Nubia, the region between the First and Second Cataracts, copper and gold items first appeared in graves of the Middle A-Group, which are dated from ca. 3600–3300 cal BCE and by 3000 cal BC copper beads, awls, and pins were reaching as far south as the Third Cataract (Edwards 2004). The earliest evidence of the production of metals in Nubia is from Old Kingdom context (ca. 2600 BC) at Buhen (Emery 1963) (Fig. 4.4) and within the temple precinct further upstream at Kerma, in contexts dated by radiocarbon to 2200–2000 cal BC (Bonnet 1986). At Buhen, the copper carbonate malachite was smelted to metallic copper using charcoal from acacia trees. The later Kerma furnace is a rectangular platform, originally covered by a vault, and heated from below. This was used for melting bronze in crucibles.

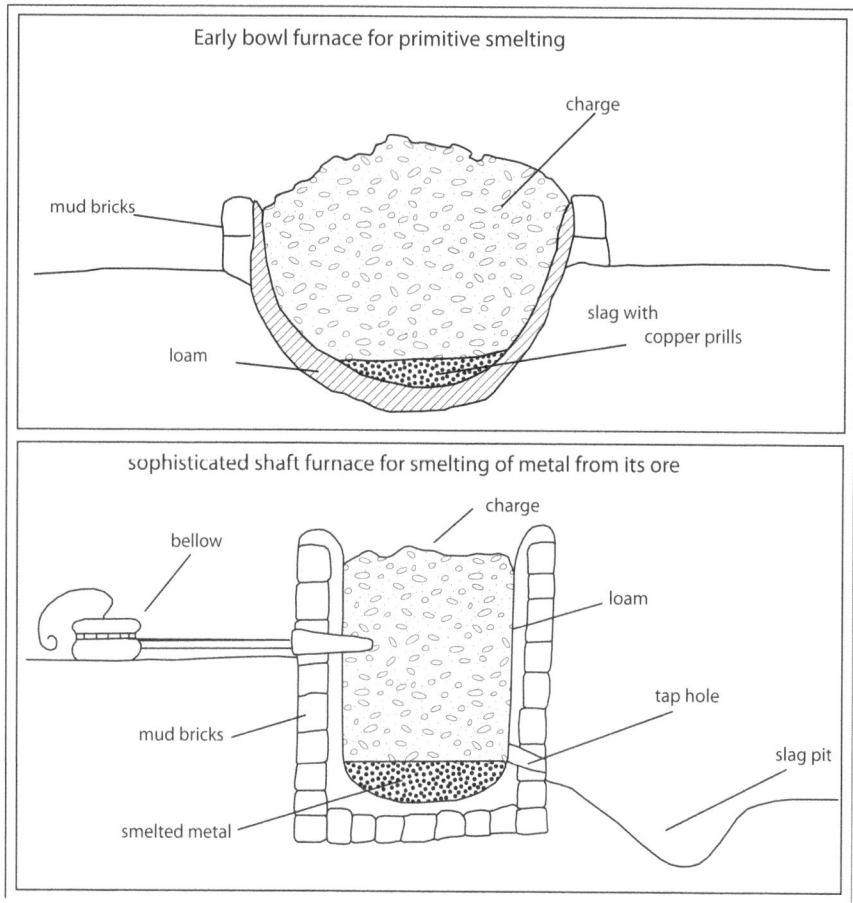

Fig. 4.7 Bowl and shaft furnaces used in dynastic Egypt. (Redrawn from Scheel 1989, p. 16, Figs. 8 and 9)

No evidence of smelting has yet been found at Kerma, and the sources of the copper and tin used are unknown (Killick 2014).

Like in Egypt, iron appeared comparatively late in Nubia with the earliest date being a radiocarbon date of 514+/−73 BC (Shinnie 1985, p. 30). Meroe, the capital of the kingdom of Kush, yielded huge quantities of smelting slag, slag, possible smithing hearths, and tuyeres (Rehren 2001). Iron production continued until the early first millennium AD when Meroe was abandoned. Meroitic furnaces were constructed of fired brick, and Shinnie (1985) found good evidence for the nature of the bellows when he exposed pot cylinders associated with furnace F5 (Fig. 4.8). It appears that Meroitic furnaces were derived from Egypt and were introduced by the Romans. Meroitic iron smelting was large scale (Fig. 4.9) and may have produced more than 5000 t of iron, which is clearly beyond local needs (Rehren 2001).

Brief Overviews: Metal Smelting in Preindustrial Africa 73

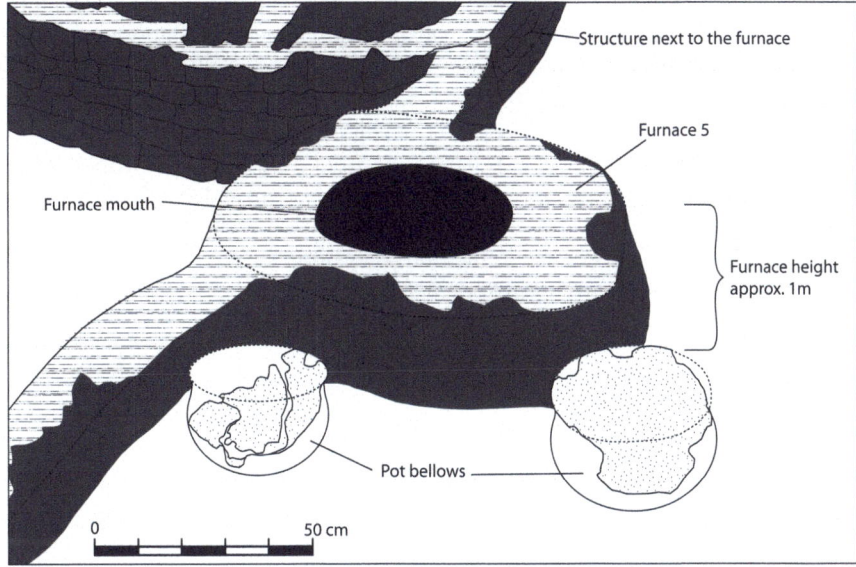

Fig. 4.8 Furnace F5 excavated by Shinnie (1985) at Meroe. (Redrawn from Shinnie 1985, p. 33)

Fig. 4.9 Large iron slag mound at Meroe. (Source: Jane Humpris)

In Ethiopia and the Horn of Africa, gold and copper appeared in the last century BC. The available evidence suggests that iron and copper metallurgy was established by c. 500 BC, possibly as a result of influence from either via Egypt and Nubia or via Arabia. The production of copper and iron was typically through the bloomery process, and Severin et al. (2011) studied the slags from Aksum dating to AD 300 and suggested that some of them were from iron smelting. The extractive metallurgy of North Africa is closely related to that of Egypt and the Mediterranean. In Morocco, the metallurgy closely mirrors that of the adjacent Spain with the object forms resembling those used in the European Bronze Age (Alpern 2005). This shows that different areas were networked and were thus not closed to each other.

West Africa

West Africa (Fig. 4.10) possesses one of the longest records of metal production in Africa south of the Sahara, excluding Nubia. Starting with the ethnographic period, a lot is known about the production of metal in West Africa (Cline 1937). Huysecom and Augustoni (1997) reconstructed iron smelting in natural draft furnaces by the Dogon of Mali encompassing the entire *chaîne opératoire* from raw material selection through smelting to actual forging. Similar reconstructions were also performed at Banjeli in Togo (Goucher and Herbert 1996). Archaeologically, detailed work was carried out in a number of areas in West Africa, which enhanced our understanding of the technology of the process within a diachronic perspective (see for example, de Barros 2013; Eze-Uzomaka 2013).

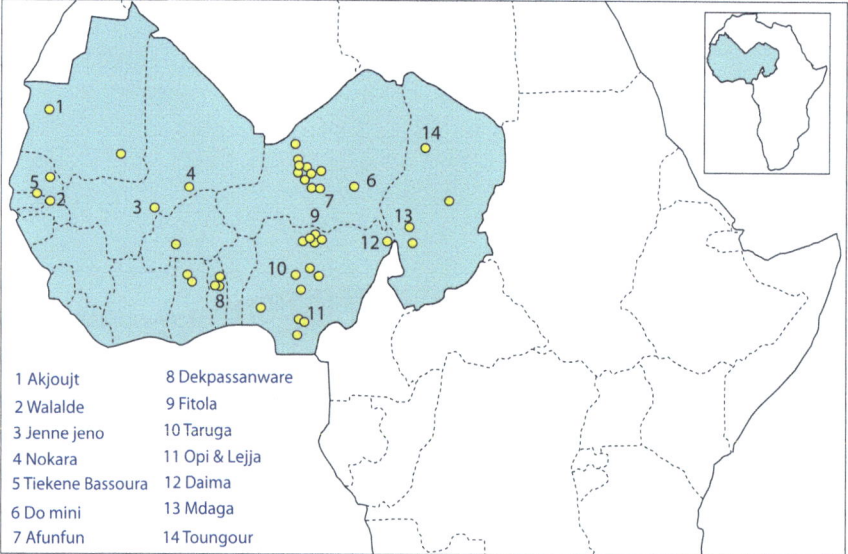

Fig. 4.10 Smelting sites in West Africa

West Africa

Fig. 4.11 Second-millennium AD furnaces used in Bassar, Togo. (Source: Philip de Barros)

De Barros (1986, 2003, 2013) carried out regional surveys in the Bassar region of Togo. In the process, he explored the development of iron production from the Early Iron Age (c. 500 BC to AD 1000) until the recent periods. The most interesting evidence comes from the multicomponent site of Dekpassanware in Bassar, Togo. Dekpassanware yielded an impressive record of smelting; often punctuated by discontinuities in the stratigraphy between c. 500 BC until recent times (Fig. 4.10). According to de Barros (2013), from c. 500 BC up to sometime in the mid-to late first millennium AD, iron smelting furnaces used at Dekpassanware were very small and were possibly powered by bellows. But from the late first millennium AD, smelting took place in natural draught furnaces (Fig. 4.11), which consumed large quantities of ore, clay and labour. As the second millennium AD unfolded, the formation of large-scale polities was associated with large-scale iron production and was characterized by division of labour with some villages specializing in ore preparation, others in smelting and some in smithing (de Barros 1986). Iron production in regions such as Banjeli and Bassar peaked up to reach up to 80,000 cubic m (Fig. 4.12), which surpasses production at places such as Meroe (de Barros 1986). No detailed studies of Early and Late Iron Age slags from Bassar have been carried to date, but analyses by Goucher indicated that fifteenth-and sixteenth-century AD slags from this area were typical bloomery waste products, while preliminary metallographic analyses of objects showed them to be made of low-carbon steels also consistent with bloomery process (de Barros 2013).

Fig. 4.12 Second-millennium AD slag mounds at Bassar, Togo. (Source: Philip de Barros)

From the late 1990s, a team led by Huysecom initiated a long-term study of iron production in the Dogon region of Mali (Robion-Brunner et al. 2013). The team was able to establish five iron smelting traditions characterized by different furnace types and scales of production within this limited area. Natural draught furnaces in this area possibly developed in the second half of the first millennium AD. The scale of production was also very high, particularly from the second millennium AD, and is comparable with that of places such as Banjeli in Togo, which were associated with increasing political centralisation. Robion-Brunner et al. (2013) speculate that the caste system may have developed in the late second millennium AD in this region.

Research in various parts of Nigeria unearthed important data which documents the development of iron working in this part of sub-Saharan Africa. In Nsukka, Okafor (1993) chronicled the evolution of iron smelting at places such as Opi dating from c. 600 BC to the late first millennium AD (Fig. 4.1). During this Early Iron Age, the furnaces were non-slag tapping. In the Late Iron Age (AD 1000–1800), tapping furnaces were used with the result that they increased output from the furnaces through reduction efficiency. This observation is supported by microstructural studies of slag from the two periods, which indicated that Late Iron Age slags had less residual iron oxide when compared to those produced in Early Iron Age furnaces (Okafor 1993). The tapped slag from Opi had macroscopically visible tap lines, while microscopically a series of magnetite skins, which develop when hot slag is exposed to cool air typically after a smelt, were detected. Elsewhere in Nigeria, Ige

Fig. 4.13 Remnants of a circular shaft furnace from an iron smelting furnace in Senegal, dated between the twelveth and fourteenth centuries AD. The black material is slag that solidified within the furnace. (Photo credit: David Killick)

and Rehren (2003) discuss the smelting of ilmenite-rich magnetite sand by the Yoruba from the eighteenth and nineteenth centuries. According to Killick and Miller (2014), such a capability showcases the versatility of the bloomery process because modern blast furnaces cannot smelt iron ores with more than 2 wt % titanium.

As these examples make clear, iron metallurgy was not static from its introduction in the first millennium BC up to the present. Sites such as Waldadé in Senegal have produced iron tools and hint of iron production on two mounds producing calibrated dates of 800–200 BC which fall in the radiocarbon black hole (Deme and McIntosh 2006) (Fig. 4.1). Earliest evidence from Waldadé indicates transitional use of iron between cal 800 and 550 BC, but the evidence shows that smelting was fully established by 200 cal BC. Studies by Killick and others in the Middle Senegal River identified many furnaces dating to the late first and early second millennium AD (Fig. 4.13) In Mauretania, there is evidence that iron was smelted between 760 and 400 cal BC (MacDonald et al. 2009). These early Mauritanian furnaces were non-slag tapping, supporting a trajectory from non-slag-tapping to tapping as recorded in areas such as Bassar. One important observation based on examples of iron production in West Africa mentioned here and elsewhere is the phenomenal boom in scale of production from the late first and early second millennium AD onwards. The volume of iron slags recorded by Robion-Brunner (et al. 2013) in the Dogon region of Mali was estimated to be in excess of 50,000 m^3 of slag, relatively comparable with the equally staggering 80,000 m^3 for Bassar and Banjeli areas of

Togo. Several reasons may account for this, but there appears to be a correlation between large-scale production and a boom in demography, mainly visible after AD 1400 (de Barros 1986); furthermore, expanding networks of connection in relation to initially trans-Saharan trade and later the Atlantic-based trade (Stahl 2014a). It is equally possible that some of the iron production may have served the demands of the slave trade (MacEachern 1993).

In contrast to the relatively robust evidence for iron working, only a few copper working localities in West Africa have been described in detail. Perhaps the most well-known example of copper working in this region seems to be at Akjoujt (850–300 BC) in Mauretania, which also represents the earliest copper south of the Sahara outside Nubia (Lambert 1983; Wood house 1998). At Azelik in Niger, copper objects were found in Late Stone Age and Iron Age contexts. The dating of the earliest appearance of copper in this region is controversial, but the main indication is that copper smelting may have started somewhere after c. 1000 BC (Killick et al. 1988). The earliest furnaces used for smelting copper were low shaft and non-slag tapping (Bisson 2000, p. 88). Copper working in Niger continued well into the second millennium AD, with interruptions in between, such that the Arab chronicler Ibn Battuta recorded copper working in this area c. AD 1300 (Herbert 1984). In Nigeria, copper was also worked in the Benue valley, which was the source of copper used in the castings at Igbo Ukwu in the late first and early second millennium AD (Chikwendu et al. 1989). Because most West African regions lacked copper, trade in this metal was the focus of interaction between communities on the two sides of the Sahara (Fenn et al. 2009).

Central Africa

Central Africa (Fig. 4.14) provides a range of insightful examples of iron and copper smelting in Africa. Warnier and Fowler's (1979) study of the Cameroonian Grassfields on the Bamenda Plateau provides one of the most staggering examples of ninettenth-century large-scale iron production. Here the scale of iron production implies mobilization of large labour inputs and quantities of ore and other raw materials resulting in a production that unmistakably was geared beyond local needs just as Bassar and Dogon discussed above. This iron production resulted in the accumulation of debris in a very short space of time estimated at 163,000 m^3 and therefore remains one of the largest examples of iron production documented in Africa (see de Barros 1986).

Research in the Mandara region of Cameroon has yielded rich insights through the work of first Renee Gardi and later Nic David and his Mandara archaeology project. Gardi recorded ethnographic cases of iron production, illustrating furnace types, elucidating embedded sociocultural metaphors and describing how the process unfolded. While Gardi's observations formed a useful archive of cultural material, David et al. (1989) persuaded one of the last Mafa smelters Dokwaza to re-enact iron smelting in a forced down draught furnace which stood to a height of

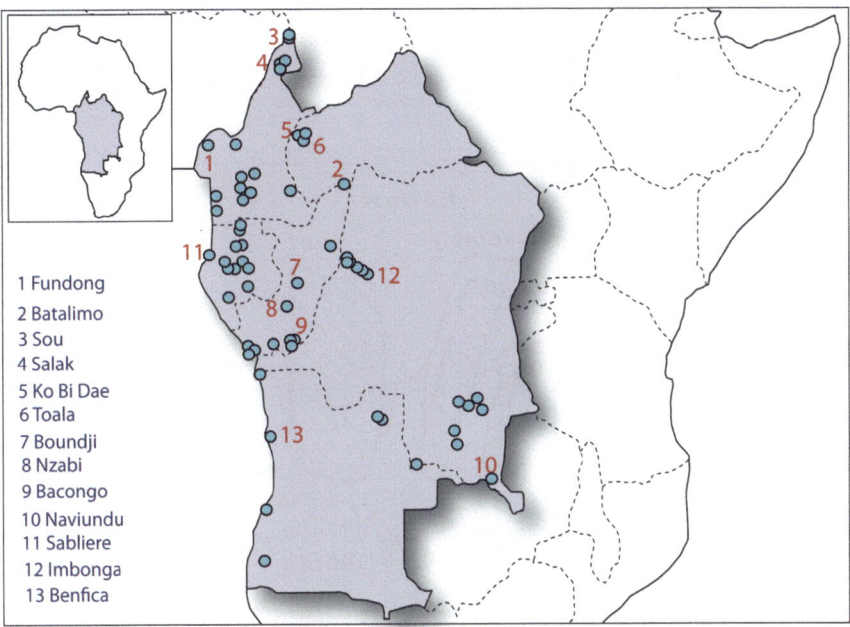

1 Fundong
2 Batalimo
3 Sou
4 Salak
5 Ko Bi Dae
6 Toala
7 Boundji
8 Nzabi
9 Bacongo
10 Naviundu
11 Sabliere
12 Imbonga
13 Benfica

Fig. 4.14 Location of Central African sites

2.7 m and was uniquely located to exploit the nature of the topography (Fig. 4.15). Two pot cylinders with an attached sheep skin diaphragm were used to generate air, which was directed vertically by a single tuyere into the furnace. Mafa smelting exploited the much iron-rich magnetite sand panned from water courses by women (Cline 1937). Over the course of smelting, the tuyere had to be trimmed of viscous slag that intermittently blocked the air hole; slag was also removed from a series of vent holes cut into the furnace wall at increasing heights. During the smelting, the tuyere lost about a third of its length (0.5 m), contributing to slag formation. After ten and half hours of nonstop smelting that consumed 82.3 kg of charcoal and 18.0 kg of ore, 15.7 kg of iron bloom was recovered (David et al. 1989). The carbon content of the bloom varied from 0.05% carbon to cast iron of more than 4% carbon. The cast iron was decarburized together with unsorted fragments of other metallic products in an open crucible. The carbon content of the final product ranged from 0.2 to 0.8%, making it a low-to high-carbon steel. It is because of this versatility and ability to produce cast iron that this Mafa iron smelting has been described as a hybrid between the bloomery and blast furnace (David et al. 1989).

Other regions of Central Africa such as Gabon have also produced important information regarding metal production (Clist 2013), but in the interest of regional balance, it is more prudent to shift focus to the Democratic Republic of Congo and adjacent areas. Most work in this region was carried out by de Maret and his students (de Maret 1988). It is clear that iron metallurgy in this region appeared much later than in Cameroon, Gabon and the Central African Republic (Clist 2013). The

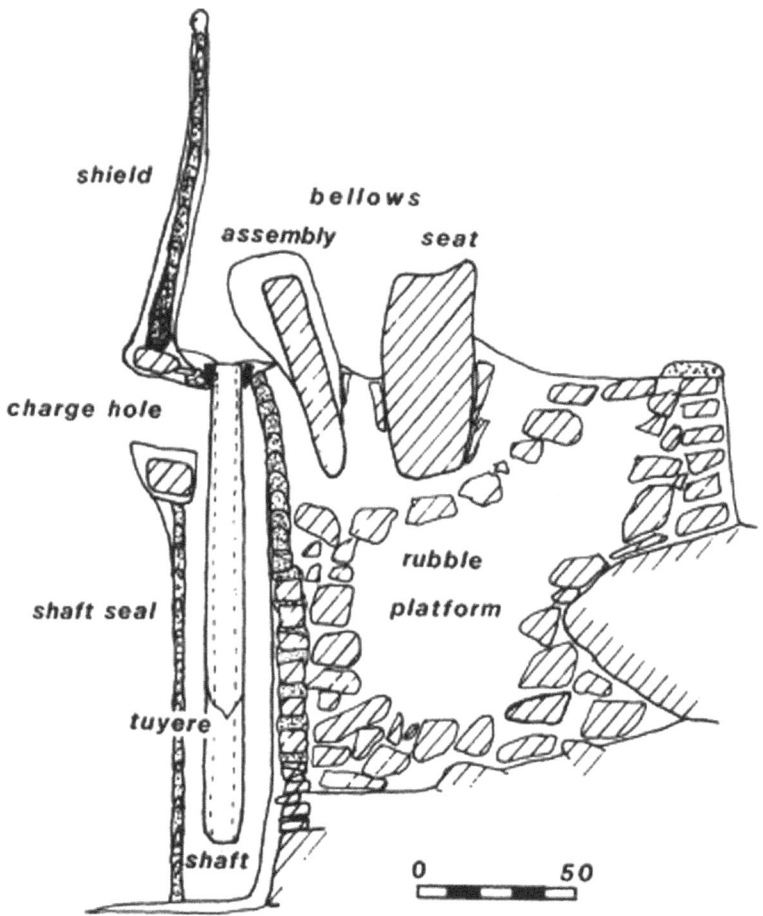

Fig. 4.15 Cross section of Mafa down draught furnace. (From David et al. 1989, p. 188, Fig. 3)

archaeological sites in the Upemba Depression yielded iron and copper objects in the early first millennium AD. As with other parts of the continent, once established, metallurgy developed over time, resulting in very rich diversity of furnace types from slag tapping bowl furnaces to low shaft and natural draught furnaces.

Although the Democratic Republic of Congo, Angola and Congo-Brazzaville have important iron production signatures, they also host significant evidence of copper production dating from the early first millennium AD to the early twentieth century. Furthermore, copper production in this area was studied within a diachronic perspective that allows us to explore in detail the development of copper smelting better than any other region in Africa (see Bisson 2000). Regions such as Lubumbashi (formerly Katanga) in the Democratic Republic of Congo (DRC), Bembe in Angola, and the copper belt in Zambia host significant ethnographic and archaeological evidence of copper smelting (Herbert 1984). This broad region in the heart of Africa possesses a rich data set of ethnographic examples of copper smelting and its associated sociological factors. According to Bisson (2000), from the nineteenth century onwards, European travelers documented copper smelting practices by groups such as the Sanga, Yeke, and Luba communities who worked in one of the largest copper deposits in the world around Lubumbashi. Ethnographically, the Sanga copper smelters (Fig. 4.1) reduced malachite in shaft furnaces that were one and three quarter meters tall and about a meter wide at the base. These bellows-driven furnaces were molded out of clay from termite mounds and had a shallow depression in the ground where molten metal collected. Upon the completion of the smelt, the furnace rake channel was opened and the solidified metal was skilfully removed without destroying the superstructure. The metal from the furnace was refined in a secondary furnace powered by four pairs of bellows to ensure that high level of heat was generated. Copper smelting was the preserve of certain kin groups, although mining could be done by both men and women of all strata (Bisson 2000).

The Luba, another group in today's DRC, had an extraordinary copper smelting technology that directly tapped molten metal into ingot molds as soon as the reduced metal reached the furnace bottom (Bisson 2000; Herbert 1984). The shaft furnace was connected to numerous but shallow X-shaped molds lined with ashes. After each mold was filled, the smelting assistants opened other vents and re-oriented the channel until all the molds were full and the furnace was empty, after which the process was started all over. This semicontinuous process resulted in the production of a significant number of crosses which were traded over wide distances.

Although the archaeological record of Central Africa is rich in evidence of copper smelting from the first millennium AD, there has been little technical study of remains from the production process. To date, the earliest evidence of copper smelting south the equator comes in the form of pieces of malachite, slag, broken pots, crucibles, heavily vitrified and slag-encrusted tuyeres, and remnants of furnaces from Naviundu Springs (Fig. 4.14) near Lubumbashi in DRC (de Maret 1982; Herbert 1984). The site dates to the third millennium AD. More evidence came from archaeological excavations at Sanga near Lake Kisale in the Upemba Depression. According to de Maret (1985), the Sanga burials represented three groups (1) Ancient Kisalian (c. eighth to tenth centuries AD), (2) Classic Kisalian (c.eleventh to fourteenth centuries AD) and (3) the Kabambian (c. fifteenth to eighteenth centuries AD), which had evidence of copper working, mostly objects. The Sanga copper funerary evidence revealed increasing social differentiation across the three periods and validated Herbert (1984)'s observation that copper was 'the red gold of Africa' and thus the prestige metal of choice.

One of the most detailed archaeological examples of preindustrial copper mining comes from Zambia at Kansanshi mine, which was worked from the early first millennium AD up to fairly recent times (Bisson 2000). Copper smelting took place away from the mine on a 150,000-square-meter-large site characterized by three major phases of occupation. Archaeological excavations by Bisson (1976) revealed that the earliest evidence for copper working designated Kansanshi Phase 1 dated to AD440+/-90. This phase yielded a crucible and a big block of slag. This was followed by Phase 2 dating from the eighth to the tenth century AD. According to Bisson (2000, p. 140), one of the furnaces used during this phase had multiple tuyeres, suggesting that it was likely powered by natural draught. This furnace differed from earlier and later ones, which had single-or two-tuyere pots indicating that they were forced draught driven. The last Phase 3 was dated between AD 1200 and 1600. The smelting debris from this area has not yet been studied in detail, but holds potential to yield insight into the operations of natural draught furnaces. These furnaces are generally believed to be heavily reducing, and in copper, smelting would have reduced more iron, which was undesirable (Craddock and Meeks 1987). It is therefore tempting to speculate that the practice of smelting copper in natural draught furnaces was abandoned for these technical reasons; however, only empirical research can validate or refute this proposition. Nonetheless, the scale of production at Kansanshi was considerable and easily reached 130 000 kg of slag for the last phase showing production beyond localized needs (Bisson 2000).

East Africa

Mapunda (1995; 2003) combined oral traditions, ethnohistories, and archaeological evidence to investigate iron production from AD 1600 up to the 1950s in Ufipa near Lake Tanganyika, South-western Tanzania. His work identified three traditions: (1) the pre-Bantu *Katukutu* or dwarf technology (c. sixteenth to eighteenth centuries AD) found on the southeastern shore of Lake Tanganyika (Mapunda 2003, p. 72); (2) the *Malungu* (tall furnaces), dating to the nineteenth and twentieth centuries AD, and located mainly on the escarpment and the Fipa plateau, and (3) the Barongo-type technology, dating to the nineteenth century and located along the lakeshore. These technologies or traditions also extended into parts of neighboring regions in countries such as DRC, Malawi and Zambia. The *Malungu* tradition was associated with the *vintengwe* furnaces used for refining blooms from the tall furnaces. Mapunda (2003, p. 82) argues that the authors of the *katukutu* tradition were of hunter–gather extraction known locally as *mbonelakuti* on the basis of linguistic, genetic and historical evidence. This, argues Mapunda, indicates that iron working technologies were not only limited to groups conventionally labeled Bantu(Fig. 4.16).

Lyaya et al. (2012) and Lyaya (2013) carried out archaeometallurgical studies combining optical microscopy with compositional techniques to understand iron production in Southern Tanzania, an area encompassing Fipa and Unyiha areas. More interestingly, Lyaya studied the slags and tuyeres from the *Malungu* and *vintengwe* furnaces to argue powerfully for a three-stage iron production process in this part of Africa. The *Malungu* process produced a sintered matrix that was further processed in the refining or *vintengwe* furnaces. The blooms from this second stage were smithed in the third and final phase of processing. According to Lyaya et al. (2012), this three-stage process differs from the two-stage (smelting and smithing) process, which generally excluded refining furnaces and was recorded in numerous African parts of sub-Saharan Africa. This indicates great variability in African bloomery processes. Lyaya also found evidence that Fipa smelters could produce

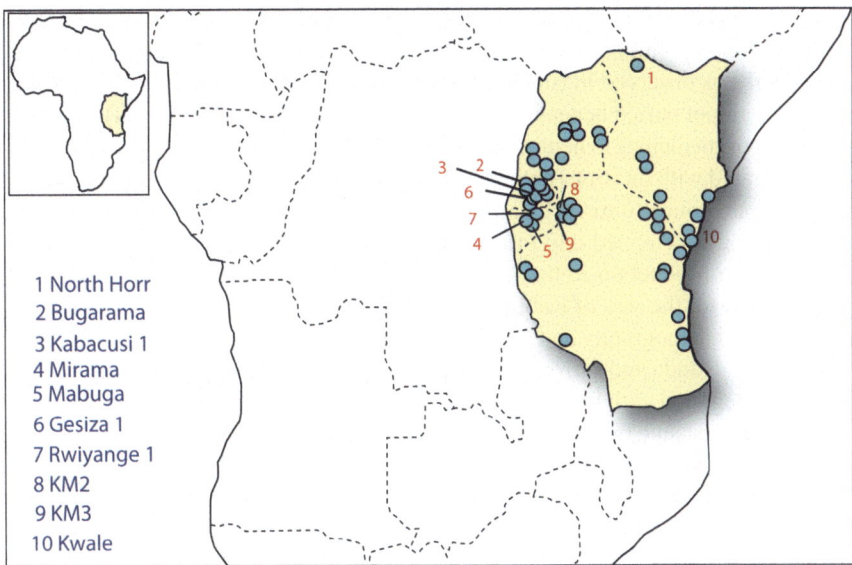

Fig. 4.16 Location of East African sites and groups

cast iron and various grades of steel, further exposing the diversity of iron production techniques across Africa.

Schmidt (1997) and collaborators carried out a detailed ethnoarchaeological study of iron production in Buhaya, Northwestern Tanzania. These studies were supplemented by nine experiments that were recorded with the aim of generating analogues for interpreting the archaeological record. The Haya people roasted their ore before smelting, used charcoal from the hard wood *Mucwezi* tree and inserted their tuyeres deep into furnace bowl with the implication that air was pre-heated in tuyeres before introduction to the furnace. These practices were combined to produce steels of variable carbon content (Schmidt 1997, p. 106–110). Schmidt and Avery (1978) argued that this preheating technology had a deep antiquity in the Iron Age, extending to between c. 500 BC and 1000 AD. In particular, the sites of KM1, KM2 and KM3 linked to the Kaiija shrine at Rugomora Mahe dating to c. 500 BC possess this evidence, though claims for preheating have been challenged (see Rehder 1986). These Early Iron Age sites were made of clay bricks, while those in the Later Iron Age were made of termite earth and slag blocks. The earliest smelting technology in Buhaya was related to that recorded in the adjacent areas of Rwanda and Burundi where Urewe pottery was recovered (Humphris and Iles 2013).

In the Interlacustrine region of East and Central Africa, van Noten (1985) re-enacted iron smelting at Madi in Northeastern Democratic Republic of Congo and the adjacent Gasara region of Rwanda. While the Madi experiment produced iron, the Gasara one failed demonstrating how colonial processes interrupted knowledge transmission. Archaeologically, Rwanda and Burundi have generated important sites crucial for our understanding of the evolution of metallurgy. The Early Iron Age in this area dates to c. 500 BC and perhaps earlier (Humphris and Iles 2013). The typical Early Iron Age furnaces were made of either bricks or coils of refractory clays. Interestingly, as observed in Buhaya Tanzania and other areas of Africa, van Noten found a small pot buried beneath the furnace pit (Van Noten 1985, p. 104), suggesting the use of medicines to neutralize malevolent forces. Some of the furnace bricks were decorated with incisions and/or dimples. The evidence suggests that the superstructure was broken to extract iron while slag accumulated in the pits. In the Late Iron Age, slag was tapped into pits outside the furnaces. Van Noten (1985) speculated that iron production in the Early Iron Age of Rwanda had strong ecological consequences for the area of Kabuye was deforested. In the mid-second millennium AD, increased demography and political centralization precipitated large-scale iron smelting in Buganda and surrounding areas. In fact, concepts of power, fertility and iron were integrated with ideas of smith-kings as was practiced in neighboring areas of DRC, Rwanda and Tanzania (Humphris and Iles 2013, p. 59).

Iles and Martinón-Torrcs (2009) studied iron production by pastoralist groups in Kenya and noted that they used bowl furnaces. The pastoral groups placed less value on copper (Bisson 2000). Overall, the evidence for copper smelting is very rare, but copper objects were recovered around Kalambo Falls in Tanzania and other areas (Mapunda 2003).

Southern Africa

Southern Africa (Fig. 4.17) presents yet another story of diversity as far as the working of metals is concerned. A number of studies aimed at understanding iron production were carried out in different parts of the subcontinent. Dewey (1991) organized re-enactments of Njanja iron smelting at Ranga in Central Zimbabwe. He persuaded the descendants of Zinwamhanga, the head smelter who features prominently in MacKenzie's (1975) study to re-enact traditional iron smelting. Zinwamhanga had also carried out a series of re-enactments in low shaft furnaces decorated with female anatomical features at the Museum of Human Sciences (formerly Queen Victoria Memorial Museum). Dewey's film of the re-enactment shows Njanja women singing and dancing together with men to entertain and perhaps help bellows operators to maintain their rhythm. While Njanja is historically depicted as a center for specialized iron production, not many large mounds of slag comparable with those recorded in West and Central Africa were recorded.

Chirikure (2006) carried out interviews with Ranga people and carried out archaeological work around Hwedza Mountains, but failed to locate large mounds comparable with those recorded in West and Central Africa. His study of Njanja smelting remains documented a very efficient reduction technology which left little residual iron oxide in the slag.

Maggs (1992) discusses a very interesting story of large-scale iron production using bowl furnaces within the context of the nineteenth-century historical Zulu state. According to Maggs (1992), Shaka Zulu reduced the once-independent specialist

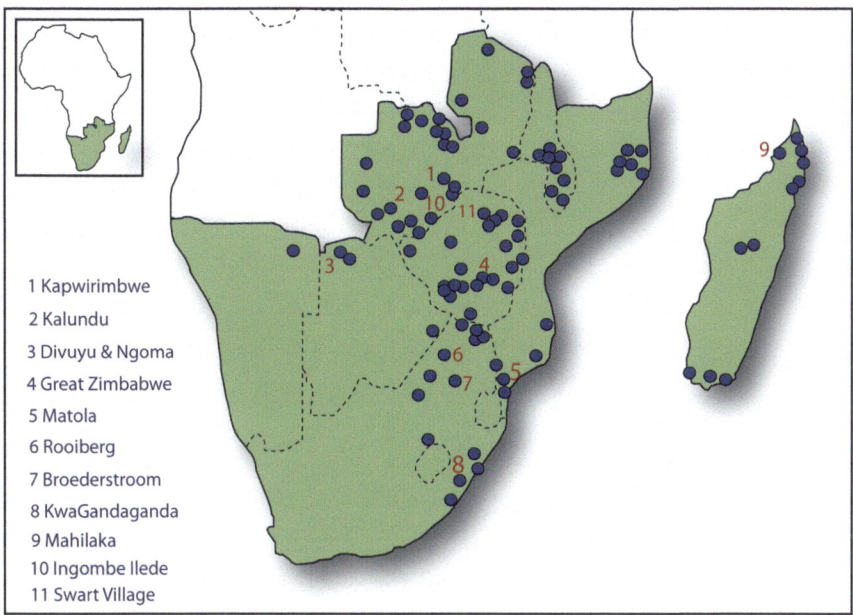

Fig. 4.17 Location of Southern African sites and groups

metal producers into dependent specialists who made iron to meet the requirements of his ever swelling army. These specialists produced iron in small bowl furnaces that fully equipped Shaka's marauding armies numbering in excess of 50,000. Interestingly, although field research may alter this supposition, no large-scale slag mounds have been noted in this area, consistent with the oral evidence that substantive production was achieved through many small-scale installations. The implication for global archaeology is that substantive outputs (large-scale production) do not necessarily correlate with concentrated production centers. Production can still be dispersed but still specialized in orientation.

There is evidence that, once established in Southern Africa, iron production developed in both versatility and complexity. It seems that bowl furnaces and shaft furnaces were used north and south of the Limpopo. Bowl furnaces were recorded in the Early Iron Age of KwaZulu-Natal (between AD 300 and 1000) and afterwards (Fig. 4.18). In other areas, such as Phalaborwa, different types of shaft furnaces were used in the Early and Late Iron Ages, showing the development of iron and copper smelting. While it is notoriously difficult to use furnaces as a proxy for group identity, Miller et al. (2001) present interesting results which associated various groups in the Northern Lowveld with specific types of furnaces particularly after AD 1400. The Venda were linked with cylindrical furnaces, while the Phabalorwa are associated with the triangular furnace. However, there are many dangers associated with conflating recent observations into the deep past.

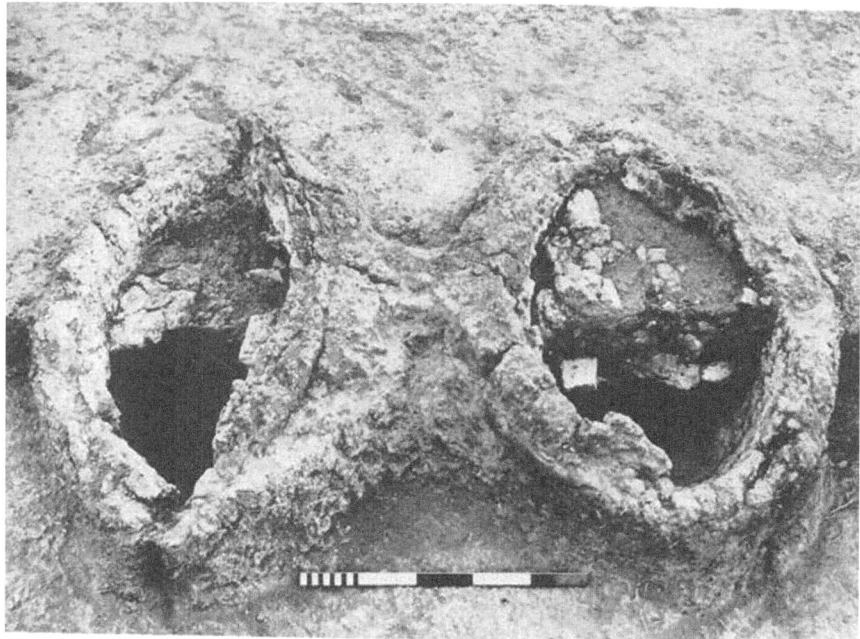

Fig. 4.18 Twin-bowl furnaces used in the Later Iron Age (c. AD 1700 onward) of KwaZulu-Natal, South Africa. (Photo credit: Tim Maggs)

Ndoro (1994) studied iron smelting remains recovered at Chigaramboni near Great Zimbabwe and concluded that, because tuyeres fused in multiples, the furnaces were powered by natural draught. No scientific studies have been performed on this material, but it is possible that this site may have supplied iron to the inhabitants of Great Zimbabwe (Ndoro 1994). Prendergast (1975) excavated a furnace with tuyeres fused in multiples dating to the fourteenth and fifteenth centuries near Darwendale in Zimbabwe. This is one of the earliest known natural draught furnaces south of the Zambezi. In the Tswapong Hills of Botswana, there exist tuyeres fused in multiples, which motivated Kiyaga-Mlindwa (1993) to argue that these late first millennium AD sites were naturally draught driven, but more work may be required to substantiate this. Interestingly, ethnographic evidence of natural draught furnaces is mostly restricted to areas north of the Zambezi, suggesting complex patterns of continuity and change in need of more substantive investigation.

Owing to the work of Miller and others before and after him, Southern Africa has witnessed a number of archaeometallurgical studies that enhanced our understanding of iron, copper and bronze technology as reconstructed from the production remains. The work of Miller and Killick (2004) around Phalaborwa identified a very versatile technology, which thrived on exploiting very high-grade but titanium-rich magnetite. It is possible that sand was added as a flux to reduce these high-grade ores. Chirikure and Rehren (2006) analyzed the archaeometallurgical remains from

Swart Village near Mt Darwin in Northern Zimbabwe. They too documented a very efficient iron reduction practice in the region.

Copper smelting is less well investigated in part because it is difficult to distinguish copper from iron smelting slags visually and in some cases chemically (Miller and Killick 2004). One of the most detailed works on archaeological copper smelting slags from Southern African sites was carried out by Thondhlana (2012) at Phalaborwa as part of a project directed by the present author. Thondhlana excavated a midden and a slag mound at the site of Shankare and identified both copper and iron smelting dating to Kgopolwe (c. 1100 to 1300) and Letaba (c. 1700 to 1900). Archaeometallurgical analyses revealed that while iron smelting slags from the site were enriched in titanium, copper slags lacked this element. Furthermore, in contrast to the wustite-dominated iron smelting slags, the copper slags had magnetite spinels and small copper prills. The copper from Shankare furnaces was melted in ceramic crucibles visually indistinguishable from normal pottery. Petrographic work on the crucibles revealed that some of them were tempered with slag inclusions identifiable through the presence of wustite and the olivine fayalite (Thondhlana 2012). Based on these interesting results, ongoing work as part of Abigail Moffet's PhD thesis at the University of Cape Town is exploring similarities and differences between Kgopolwe and Letaba iron and copper smelting at the site to determine whether the two metals were associated with similar or different technological styles. Without imposing any relationships, cultural or otherwise, slag tempering of pottery has been noted at archaeological sites of various periods in Banda, modern day Ghana (Stahl pers comm 2014).

Anthropology of Smelting

One of the most important topics in African metal smelting relates to the symbolic and cultural aspects of the process, which clearly were not only restricted to this process but extended to broader society (Chirikure 2007). A number of important works, for example, Herbert (1993), have articulated the inextricable nature of processes such as iron smelting and rituals of power, gender and transformation across sub-Saharan Africa. In fact, as a heat-mediated routine, smelting transforms objects of nature into culture. This act of transformation is often conceived to be a dangerous process, with implications for the social position of smelters which varied geographically and through time. In some societies, particularly as documented in the ethnography of Central Africa, smelters were held in awe, while in others such as Ethiopia, they were despised. Smelters in Mali and Ethiopia formed an endogamous caste that only married potters and were of a lowly social position (Haaland 2004; Tamari 1991). At the same time, smelters and smiths could occupy privileged positions through which they accumulated wealth in Central Africa (de Maret 1985). Regardless of variation, smelters were associated with beliefs in magic and deities across the continent but in different contexts. In light of the power of evil spirits to influence smelts, smelters were motivated to place medicines beneath furnaces.

As such, some furnaces in many parts of Africa, from the Grasslands of Cameroon, through Central Africa to Eastern and Southern Africa, both ethnographic and archaeological, had medicine holes strategically placed underneath (Killick 2014; Mapunda 1995; Rowlands and Warnier 1993; Schmidt and Mapunda 1995).

One of the most important themes in the anthropology of iron smelting, particularly in Bantu Africa, is the metaphoric association between reduction and human reproduction and the implications for spatial location of smelting in relation to settlements (Ndoro 1991). Iron smelting furnaces in East and Southern Africa are often decorated with female anatomical features such as breasts (Figs. 4.19 and 4.20). Some ethnographies argue that, because of its link with reproduction, smelting took place in secluded areas away from settlements (Van der Merwe and Avery 1987). This was partly meant to enforce sexual abstinence, for intercourse with their real wives would be tantamount to adultery which gestated failed smelts (Herbert 1993). Indeed, there are some furnaces such as those in the Matopos region of Southwestern Zimbabwe and Nyanga which were located away from settlements (Soper 2002). However, most of the studies on this topic have lacked some nuance and variation because Hatton (1967) has reported a case in nineteenth-century Zimbabwe where furnaces were often located in homesteads with women and children watching. Hatton (1967, p. 39) argues that 'often, a wife would help her husband by pumping the bellows'. This variation contradicts some ethnography that sees no role for women in smelting. The Kalanga case described by Hatton (1967) exemplifies improvisation in preindustrial African metalworking (cf. Stahl 2014b). This improvisation enabled communities to transcend the usual and is often missed by an adherence to sweeping generalisations. This tempers the view that in Southern and Eastern Africa from the beginning of metallurgy until recent times, smelting was always carried outside settlements, regardless of time and place because of its association with metaphors of reproduction and concepts of pollution.

In fact, there are numerous areas in the recent past where smelting was carried out either in relatively secluded areas within the centre of villages or in full view of everybody (MacKenzie 1975; Chirikure unpublished field notes). My own field research carried out in the Njanja area reached concordance with Hatton's observations in the Matopos. My informant Headman Ranga categorically stated that smelting was carried out in the villages because the labour of women and children was important in times of high demand. Furthermore, Njanja women also sang and danced to help bellow workers maintain rhythm (Dewey 1991). Archaeologically, there are well-documented examples where smelting took place near or within villages as well as outside villages from the distant to the near past. Generally, the indicators of smelting are partially reduced ore, flow slag, remnants of furnaces, and vitrified tuyeres (Miller and Killick 2004). These cannot be mistaken for smithing for the two processes are outwardly different (Bachmann 1982; Serneels and Perret 2003).

An intact furnace surrounded by a heap of slag mixed with broken tuyeres, vitrified furnace wall and partially reduced ore was found at a nineteenth-century site in upland Nyanga, Eastern Zimbabwe, in association with a decorated furnace with breasts and a waist belt (Fig. 4.19) (Chirikure and Rehren 2004; Soper 2002). This

Fig. 4.19 Anthropomorphic low shaft iron smelting furnace from Nyanga, Eastern Zimbabwe. Note the molded breasts, navel and waist belt for enhancing fertility. (Photo credit: Author)

furnace was in a low enclosure within the homestead precinct as indicated by pit structures and house floors (Chirikure and Rehren 2004). The ceramics found on the metal smelting area of the settlement were the same as those from the house floors indicating some contemporaneity. This example demonstrates that metaphors of reproduction were important in this smelting as shown by the elaborately decorated furnace (Fig. 4.19). Indeed, the low enclosure may have had a practical function of demarcating space, but equally it may have been part of a structure which shielded smelters from general view and was therefore an expression of symbolic

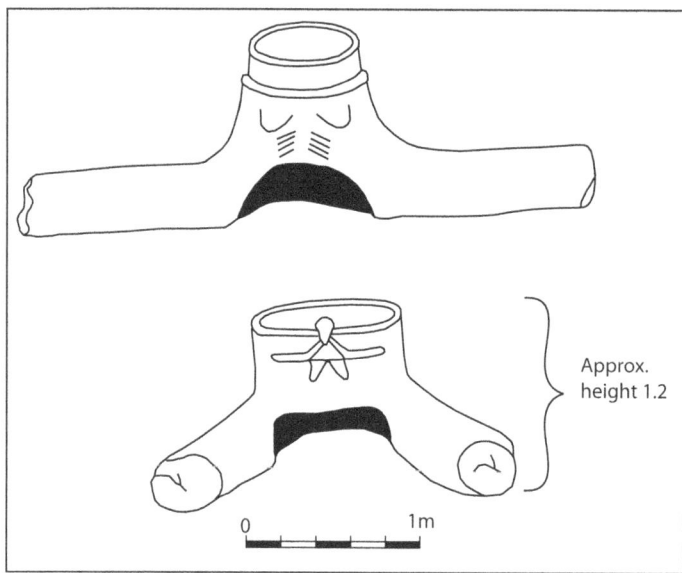

Fig. 4.20 Decorated furnaces, one depicting a woman giving birth. Redrawn from furnaces on display at Natural History Museum, Bulawayo

seclusion. Many other nineteenth-century furnaces in lowland Nyanga were located away from homesteads (Bernhard 1962; Chirikure and Rehren 2004). This variation in the spatial location of smelting in relation to settlements shows that it is not the physicality of the location that matters, but an adherence to cultural principles and values. As such, it may not have mattered whether a furnace was contiguous to, or far from a settlement, at the time as smelters as actors were always producing and reproducing ideas prevalent in society. Archaeological evidence therefore has great potential to illuminate not only various trajectories of the spatial and symbolic association between smelting and settlements but also temporal improvisation and change in those ideas and values (see Stahl 2014b).

This thinking is further substantiated by the observations made by Bent (1896) who recorded items of material culture such as houses, granaries, drums and head rests decorated with anatomical features such as breasts, navels and other fertility iconography (Fig. 4.21). All these were within the context of the homestead, and the decorations resembled those on furnaces situated away from settlements. If fertility symbolism was so important in determining the remote location of furnaces together with the associated taboos, why then were houses and other household material culture similarly decorated? (see also Collett 1993) Instead, we should consider metal smelting as an integral element of society which also produced and reproduced ideas that pervaded society such as fertility. Therefore, whether smelting was within a village or outside may be immaterial; what is material is that the process produced and reproduced ideas associated with reproduction, witchcraft and so on. Equally, we should recognize that insights from ethnographically documented

Anthropology of Smelting 91

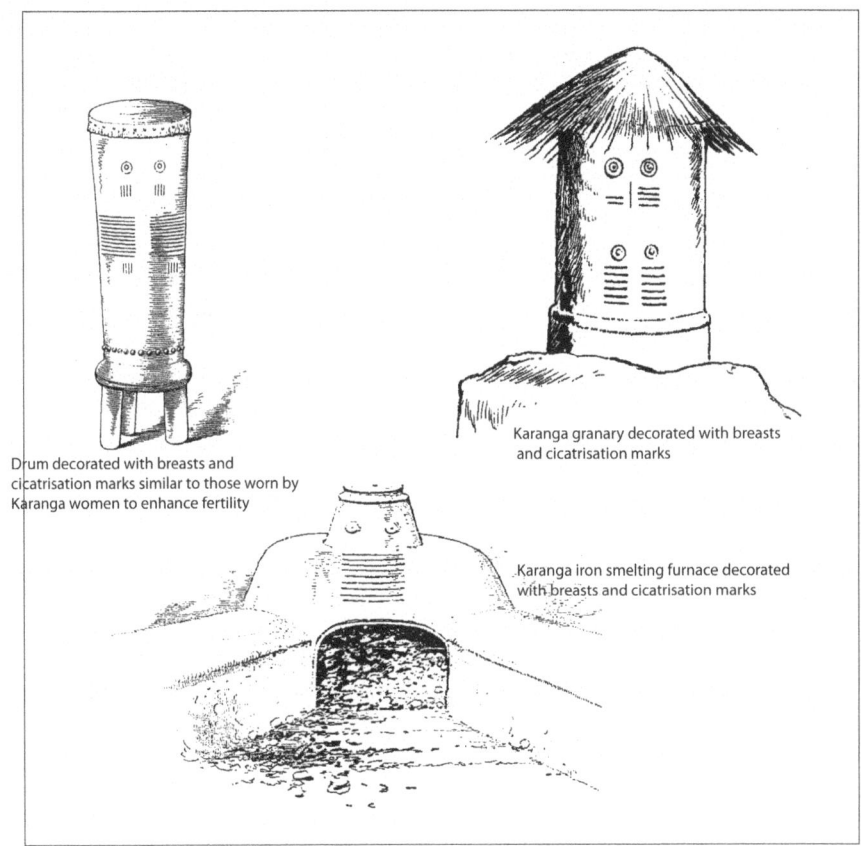

Fig. 4.21 Anthropomorphic drum, granary and iron smelting furnace from Bent (1896)

practices of the recent past may not provide a reliable guide to past variation. At the same time, it should be noted that it was not only cultural factors that determined the location of furnaces but also more practical ones such as availability of ore and fuel and also labour (Maggs 1982).

Archaeological evidence provides a key means for probing continuities and change in practice, including the relationship between smelting and residential space. Schmidt (1997) excavated Early Iron Age sites in Buhaya, Tanzania, where house floors contiguous to iron smelting furnaces were found at both Rugomora Mahe and KM sites in Northwestern Tanzania. Schmidt (1997, p. 17) categorically stated that *'I want to be cautious that a false dichotomy is not drawn between settled life and iron production. Such a dichotomy appears to be based partly on the hidden inference that iron smelting would have been conducted outside village precincts, an idea that is drawn from inappropriate and incomplete ethnographic analogy: that iron smelting is always conducted in secret outside of villages. In fact, iron smelting was sometimes conducted within village precincts (cf. Killick 1990) and in the case of recent Haya smelting, in a zone contiguous to village precincts.... Simple logic also informs us that people must have lived nearby their places of work'.*

Indeed, Haaland (1994) also excavated iron smelting debris within the center of a settlement at Dakawa in Tanzania. In Southern Africa, sites such as Swart Village have yielded smelting evidence from the center and precincts of the village together with many more other examples. It is now widely accepted that the Early Iron Age in Eastern and Southern Africa is closely related (Huffman 2007; Phillipson 2005), such that it is reasonable to infer that we are dealing with ancestral Bantu peoples. That smelting within villages has been documented at related sites in the north (East Africa) and in some places such as Swart Village in Northern Zimbabwe should be enough evidence to warn archaeologists against the 'tyranny' of selective use of the ethnography (Wobst 1978). The inference does not stand that if smelting took place in villages, then it was not associated with metaphors of reproduction. Equally, it is inappropriate to deny any involvement of women in smelting. They cooked and brought food to the smelters; they often participated in mining, charcoal preparation and in some cases even took part in smelting (Dewey 1991). In Lubumbashi, one woman was very famous for leading delegations that included men who extracted copper ore (Bisson 2000). This variation indicates that there is no reason to select only one aspect of the ethnography and extrapolate it to the deeper past to the exclusion of other patterns that may—or may not—be documented in ethnographic sources.

In ancient Egypt, it has been argued that women were generally spared smelting and metalworking because they are heavy tasks (Scheel 1989). However, most smelters were linked to temples, and deities performed an important role in ensuring success.

Conclusion

In conclusion, not only is Africa characterized by a great diversity in metals worked across regions, but evidence suggests that methods developed in different ways between the various regions. Egypt, Nubia, North Africa and the Horn of Africa seem to mirror a similar trajectory. Egypt's copper working history demonstrates diverse approaches, from use of crucibles to shaft furnaces, and introducing air through simple blowing by mouth to use of varieties of pot bellows. In contrast, West, Central, East and Southern Africa show greater diversity and innovation in iron working with furnaces appearing in different forms. The scale of production clearly increased through time with varying output in West, East and Central Africa. The output in areas of Central Africa even surpasses that of Meroe by rough calculation (Warnier and Fowler 1979). It seems that demography played an important role in increased scale because Southern Africa appears to have relatively low populations when compared to East and West Africa. As such, no large-scale mounds of slag have been reported even in landscapes historically associated with specialization such as Njanja. The variation in Africa's iron smelting furnaces is matched by the products which range from soft iron, to cast iron and steel, which can only be documented through detailed metallurgical analyses.

Besides this technological diversity, metal smelting was associated with rituals, deities and beliefs. It is absolutely critical to avoid 'selective use of the ethnography' (Lane 2005) by searching for macro-and micro-variation. For example, women may not have played a direct role in the actual process of smelting in some communities (Hatton 1967), but they prepared food for smelters and in some cases even sang during smelting (Dewey 1991). The spatial location of smelting varied and was determined by diverse variables including location of ore, fuel, clay and so on. Wherever smelting took place, it produced and reproduced ideas in society; thus, we should avoid assuming that smelting that took place outside villages was more ritually charged than smelting practiced within villages.

References

Ackerman, K. J., Killick, D. J., Herbert, E. W., & Kriger, C. (1999). A study of iron smelting at Lopanzo, Equateur Province, Zaire. *Journal of Archaeological Science, 26*(8), 1135–1143.

Alpern, S. B. (2005) Did they or didn't they invent it? Iron in sub-Saharan Africa. *History in Africa, 32,* 41–94.

Arkell, A. J. (1966). The iron age in the Sudan. *Current Anthropology, 7*(4), 451–452.

Bachmann, H. (1982). *The Identification of slags from archaeological sites.* Occasional Publication no 6. London: Institute of Archaeology.

Bayley, J., & Rehren, T. (2007). Towards a functional and typological classification of crucibles. In S. La Niece, D. Hook, & P. Craddock (Eds.), *Metals and mines: Studies in archaeometallurgy* (pp. 46–55). London: Archetype.

Bent, J. T. (1896). *The ruined cities of Mashonaland: Being a record of excavation and exploration in 1891.* London: Longmans, Green, and Co.

Bernhard, F. O. (1962). Two types of iron smelting furnaces on Ziwa Farm (Inyanga). *South African Archaeological Bulletin, 17,* 235–236.

Bisson, M. S. (1976). *The prehistoric copper mines of Zambia.* Ann Arbor, Michigan: University Microfilms International.

Bisson, M. (2000). Precolonial copper metallurgy: sociopolitical context. In M. Bisson, P. de Barros, T. C. Childs, & A. F. C. Holl (Eds.), *Ancient African metallurgy: The sociocultural context* (pp. 83–146). Walnut Creek: AltaMira.

Bonnet, C. (1986). Un atelier de bronziers Ã Kerma. In M. Krause (Ed.), Nubische Studien: Tagungsakten der 5. Internationalen Konferenz der International Society for Nubian Studies, Heidelberg, 22.-25. September 1982 (pp. 19-23). Mainz am Rhein: von Zabern.

Chikwendu, V. E., Craddock, P. T., Farquhar, R. M., Shaw, T., & Umeji, A. C. (1989). Nigerian sources of copper, lead and tin for the Igbo-Ukwu bronzes. *Archaeometry, 31*(1), 27–36.

Childs, S. T. (1998). 'Find the *ekijunjumira*': Iron mine discovery, ownership and power among the Toro of Uganda. In A. B. Knapp, V. C. Piggott, & E. W. Herbert (Eds.), *Social approaches to an industrial past: the archaeology and anthropology of mining,* (pp. 123–137). Abingdon: Routledge.

Chirikure, S. (2005). *Iron production in Iron Age Zimbabwe: stagnation or innovation.* Unpublished Doctoral dissertation, University College London.

Chirikure, S. (2006). New light on njanja iron working: Towards a systematic encounter between ethnohistory and archaeometallurgy. *South African Archaeological Bulletin, 61,* 142–151.

Chirikure, S. (2007). Metals in society: Iron production and its position in Iron Age communities of Southern Africa. *Journal of Social Archaeology, 7*(1), 72–100.

Chirikure, S., & Bandama, F. (2014). Indigenous African furnace types and slag composition—Is there a correlation? *Archaeometry, 56*(2), 296–312.

Chirikure, S., & Rehren, T. (2004). Ores, furnaces, slags, and prehistoric societies: aspects of iron working in the nyanga agricultural complex, AD 1300–1900. *African Archaeological Review, 21*(3), 135–152.

Chirikure, S., & Rehren, T. (2006). Iron smelting in pre-colonial Zimbabwe: Evidence for diachronic change from Swart village and Baranda, Northern Zimbabwe. *Journal of African Archaeology, 4,* 37–54.

Chirikure, S., Burrett, R., & Heimann, R. B. (2009). Beyond furnaces and slags: a review study of bellows and their role in indigenous African metallurgical processes. *Azania: Archaeological Research in Africa, 44*(2), 195–215.

Cline, W. B. (1937). *Mining and metallurgy in Negro Africa* (No. 5). Wisconsin: George Banta.

Clist, B. (2013). Our iron smelting 14C dates from Central Africa: From plain appointment to a full blown relationship. In J. Humpris and Th. Rehren (Eds.), *The world of iron* (pp. 22–28). London: Archetype.

Collett, D. P. (1993). Metaphors and representations associated with precolonial iron- smelting in eastern and southern Africa. In T. Shaw, P. Sinclair, B. Andah, & A. Okpoko (Eds.), *The archaeology of Africa: food, metals and towns* (pp. 499–511). London: Routledge.

Craddock, P. T. (1995). *Early metal mining and production*. Washington, DC: Smithsonian University Press.

Craddock, P. T. (2000). From hearth to furnace: Evidences for the earliest metal smelting technologies in the Eastern Mediterranean. *Paléorient, 26(*2), 151–165.

Craddock, P. T., & Meeks, N. D. (1987). Iron in ancient copper. *Archaeometry, 29*(2), 187–204.

Crew, P. (1991). The experimental production of prehistoric bar iron. *Historical Metallurgy, 25*(1), 21–36.

David, N., Heimann, R., Killick, D., & Wayman, M. (1989). Between bloomery and blast furnace: Mafa iron-smelting technology in North Cameroon. *African Archaeological Review, 7*(1), 183–208.

de Barros, P. (1986). Bassar: A quantified, chronologically controlled, regional approach to a traditional iron production centre in West Africa. *Africa, 56*(2), 152–174.

de Barros, P. (1988). Societal repercussions of the rise of large-scale traditional iron production: A West African example. *African Archaeological Review, 6*(1), 91–113.

de Barros, P. (2013) A comparison of early and later iron age societies in the Bassar region of Togo. In J. Humpris & Th. Rehren (Eds.), *The world of iron* (pp. 10–21). London: Archetype.

De Maret, P. (1982). New survey of archaeological research and dates for West-Central and North-Central Africa. *Journal of African History, 23,* 1–15.

de Maret, P. (1985). The smith's myth and the origin of leadership in Central Africa. In R. Haaland & P. Shinnie (Eds.), *African iron working—ancient and traditional* (pp. 73–87). Oslo: Norwegian University Press.

Deme, A., & McIntosh, S. K. (2006). Excavations at Walaldé: New light on the settlement of the middle senegal valley by iron-using peoples. *Journal of African Archaeology, 4*(2), 317–347.

Dewey, W. J. (1991). *Weapons for the ancestors*. Des Moines, University of Iowa, Department of Art History.

Duell, P. (1938). The Mastaba of Mereruka. University of Chicago Oriental Institute.

Edwards, D. N. (2004). *The Nubian past: An archaeology of the Sudan*. London: Routledge.

El-Rahman, Y. A., Surour, A. A., El Manawi, A. H. W., Rifai, M., Motelib, A. A., Ali, W. K., & El Dougdoug, A. M. (2013). Ancient mining and smelting activities in the wadi abu gerida area, central eastern desert, Egypt: Preliminary results. *Archaeometry, 55*(6), 1067–1087.

Ellert, H. (1993). *Rivers of gold*. Gweru: Mambo Press.

Emery, W. B. (1963). Egypt exploration society preliminary report on the excavations at Buhen, 1962. *Kush, 11,* 116–120.

Emery, W. B. (1971). Preliminary report on the excavations at North Saqqâra, 1969-70. *The Journal of Egyptian Archaeology, 57,* 3–13.

References

Eze-Uzomaka, P. (2013). Iron and its influence on the prehistoric site of Leija. In J, Humpris & Th. Rehren (Eds.), *The world of iron* (pp. 3–9). London: Archetype.

Fenn, T. R., Killick, D. J., Chesley, J., Magnavita, S., & Ruiz, J. (2009). Contacts between West Africa and Roman North Africa: Archaeometallurgical results from Kissi, northeastern Burkina Faso. In S. Magnavita, L. Koté, P. Breunig, and O. A. Idé (Eds.), *Crossroads/Carrefour Sahel: Cultural and technological developments in first millennium BC/AD West Africa* (pp. 119–146). Journal of African Archaeology Monograph Series 2. Frankfurt am Main: Africa Magna Verlag.

Goucher, C. L. (1981). Iron is iron'til it is rust: trade and ecology in the decline of West African iron-smelting. *The Journal of African History, 22*(02), 179–189.

Goucher, C., & Herbert, E. (1996). The blooms of Banjeli: Technology and gender in West African iron making. In P. R. Schmidt (Ed.), *The culture and technology of African iron production* (pp. 40–57). Gainesville: University of Florida Press.

Haaland, R. (1980). Man's role in the changing habitat of mema during the old kingdom of Ghana. *Norwegian Archaeological Review, 13*(1), 31–46.

Haaland, R. (1994). Dakawa: An early Iron Age site in the Tanzanian hinterland. *Azania, 29*(1), 238–247.

Haaland, R. (2004). Iron smelting—a vanishing tradition: ethnographic study of this craft in southwest Ethiopia. *Journal of African Archaeology, 2* (1), 65–79.

Hatton, J. S. (1967). Notes on Makalanga iron smelting. *Nada, 9*(4), 39–42.

Hauptmann A. (2007). *The archaeometallurgy of copper: Evidence from Faynan, Jordan*. New York: Springer.

Heimann, R. B., Chirikure, S., & Killick, D. (2010). Mineralogical study of precolonial (1650–1850 CE) tin smelting slags from Rooiberg, Limpopo Province, South Africa. *European Journal of Mineralogy, 22*(5), 751–761.

Herbert, E. W. (1984). *Red gold of Africa: Copper in precolonial history and culture*. Madison: University of Wisconsin Press.

Herbert, E. W. (1993). *Iron, gender, and power: Rituals of transformation in African societies*. Bloomington: Indiana University Press.

Horne, L. (1982). Fuel for the metal worker. The role of charcoal and charcoal production in ancient metallurgy. *Expedition, 25,* 6–13

Huffman, T. N. (2007). *Handbook to the Iron Age: The archaeology of pre-colonial farming societies in Southern Africa*. Durban: University of Kwazulu-Natal Press.

Humpris, J. and Iles, L. (2013). Pre-colonial iron production in great lakes Africa: recent research at UCL institute of archaeology. In J. Humpris & Th. Rehren (Eds.), *The world of iron* (pp. 56–64). London: Archetype.

Huysecom E., & Agustoni B. (1997). *Inagina: l'ultime maison du fer/the last house of iron*. Geneva: Telev. Suisse Romande. Videocassette, 54 min.

Ige, A., & Rehren, Th. (2003). Black sand and iron stone: iron smelting in Modakeke, Ife, Southwestern Nigeria. *Institute for Archaeo-Metallurgical Studies, 23,* 15–20.

Iles, L., & Martinón-Torres, M. (2009). Pastoralist iron production on the Laikipia Plateau, Kenya: wider implications for archaeometallurgical studies. *Journal of Archaeological Science, 36*(10), 2314–2326.

Joosten, I., Jansen, J. B. H., & Kars, H. (1998). Geochemistry and the past: estimation of the output of a Germanic iron production site in the Netherlands. *Journal of Geochemical Exploration, 62*(1), 129–137.

Kense, F. (1985). The initial diffusion of iron to Africa. In R. Haaland & P. L. Shinnie (Eds.), *African Ironworking–ancient andtraditional* (pp. 11–35). Oslo: Norwegian University Press.

Killick, D. J. (1990). Technology in Its Social Setting: Bloomery Iron-smelting at Kasunga, Malawi, *1860-1940*. University Microfilms.

Killick D. (1991). A little-known extractive process: Iron smelting in natural-draft furnaces. *JOM, 43*(4), 62–64.

Killick, D. (2004). Review essay: What do we know about African iron working? *Journal of African Archaeology,* 2(1), 97–112.

Killick, D. (2014). Cairo to Cape: The spread of metallurgy through Eastern and Southern Africa. In B. W. Roberts & C. Thornton (Eds.), *Archaeometallurgy in global perspective* (pp. 507–527). New York: Springer.

Killick, D., & Miller, D. (2014). Smelting of magnetite and magnetite-ilmenite iron ores in the Northern Lowveld, South Africa, ca. 1000 CE to ca. 1880 CE. *Journal of Archaeological Science, 43,* 239–255.

Killick, D., Van Der Merwe, N. J., Gordon, R. B., & Grébénart, D. (1988). Reassessment of the evidence for early metallurgy in Niger, West Africa. *Journal of Archaeological Science, 15*(4), 367–394.

Kiyaga-Mulindwa, D. (1993). The iron age peoples of East–Central Botswana. In T. Shaw, P. Sinclair, B. Andah, & A. Okpoko (Eds.), *The archaeology of Africa: Food, metals, and towns* (pp. 386–390). New York: Routledge.

Lambert, N. (1983). Nouvelle contribution Ã l'étude du Chalcolithique de Mauritanie. In N. Échard (Ed.). *Métallurgies Africaines, Nouvelles contributions* (pp. 63–87). Mémoires de la Sociéte des Africanistes 9. Paris.

Lane, P. J. (2005). Barbarous tribes and unrewarding gyrations? The changing role of ethnographic imagination in African archaeology. In A. B. Stahl (Ed.), *African archaeology: A critical Introduction* (pp. 24–54). Oxford: Blackwell.

Lyaya, E. (2013). Use of charcoal species for iron working in Tanzania. In J. Humpris & Th. Rehren (Eds.), *The world of iron* (pp. 444–453). London: Archetype.

Lyaya, E.C., Mapunda, B. B., & Rehren, Th. (2012). The bloom refining technology in Ufipa, Tanzania (1850-1950). In C. Robion-Brunner & B. Martinelli (Eds.), *Metallurgie du fer et Societes africaines: Bilans et nouveaux paradigmes dans la recherche anthropoloque et archeologique* (pp. 195–207). Oxford: Archaeopress.

MacDonald, K. C., Vernet, R., Martinón-Torres, M., & Fuller, D. Q. (2009). Dhar Néma: From early agriculture to metallurgy in Southeastern Mauritania. *Azania, 44*(1), 3–48.

MacEachern, S. (1993). Selling the iron for their shackles: Wandala-Montagnard interactions in Northern Cameroon. *Journal of African History, 34,* 247–270.

Mackenzie, J. M. (1975). A pre-colonial industry: The njanja and the iron trade. *Nada, 11*(2), 200–220.

Maggs, T. (1992). 'My father's hammer never ceased its song day and night': The zulu ferrous metalworking industry. *Southern African Humanities, 4,* 65–87.

Mapunda, B. B. (1995). *An archaeological view of the history and variation of ironworking in Southwestern Tanzania*. Unpublished Doctoral dissertation, University of Florida, Gainesville.

Mapunda, B. (2003). Fipa iron technologies and their implied social history. In C. M. Kusimba & S. B. Kusimba (Eds.), *East African archaeology: Foragers, potters, smiths, and traders* (pp. 71–86). Philadelphia: University of Pennsylvania Museum Press.

Martinón-Torres, M., & Rehren, T. (2014). Technical ceramics. In B. W. Roberts & C. Thornton (Eds.), *Archaeometallurgy in global perspective: Methods and syntheses* (pp. 107–131). New York: Springer.

Miller, D., & Killick, D. (2004). Slag identification at Southern African archaeological sites. *Journal of African Archaeology, 2*(1), 23–47.

Miller, D. E., & Van Der Merwe, N. J. (1994). Early metal working in sub-Saharan Africa: A review of recent research. *Journal of African History, 35*(1), 1–36.

Miller, D., Killick, D., & Van der Merwe, N. J. (2001). Metal Working in the Northern Lowveld, South Africa AD 1000–1890. *Journal of Field Archaeology, 28*(3–4), 3–4.

Morton, G. R., & Wingrove, J. (1969). Constitution of bloomery slags: Part 1: Roman. *Journal of the Iron and Steel Institute, 207,* 1556–1564.

Ndoro, W. (1991). Why decorate her. *Zimbabwea, 3,* 5–13.

Ndoro, W. (1994). Natural draught furnaces south of the Zambezi River. *Zimbabwean Prehistory, 14,* 29–32.

Nixon, S., Rehren, T., & Guerra, M. F. (2011). New light on the early Islamic West African gold trade: coin molds from Tadmekka, Mali. *Antiquity, 85*(330), 1353–1368.

Oddy, A. (1984). Gold in the Southern African Iron Age. *Gold Bulletin, 17*(2), 70–78.

References

Ogden, J. (2000). Metals. In P. Nicholson & I. Shaw (Eds.), *Ancient Egyptian materials and technology* (pp. 148–176). Cambridge: Cambridge University Press.

Okafor, E. (1993). New evidence on early iron-smelting in Southeastern Nigeria. In T. Shaw, P. Sinclair, B. Andah, & A. Okpoko (Eds.), *The archaeology of Africa: Food, metals, and towns* (pp. 432–448). London: Routledge.

Phillipson, D. W. (2005). *African archaeology*. Cambridge: Cambridge University Press.

Pleiner, R. (2000). Iron in archaeology: the European bloomery smelters. Archeologický ústav.

Prendergast, M. D. (1975). A new furnace type from the darwendale dam basin. *Rhodesian Prehistory, 7*(14), 16–20.

Rehder, J. E. (1986). Use of preheated air in primitive furnaces: Comment on views of Avery and Schmidt. *Journal of Field Archaeology, 13*(3), 351–353.

Rehren, Th. (2001). Meroe, iron and Africa. *Der Antike Sudan, 12,* 102–109.

Rehren, T., & Pernicka, E. (2008). Coins, artefacts and isotopes–archaeometallurgy and archaeometry. *Archaeometry, 50*(2), 232–248.

Rehren, T., Charlton, M., Chirikure, S., Humphris, J., Ige, A., & Veldhuijzen, H. A. (2007). Decisions set in slag: the human factor in African iron smelting. In S. La Niece, D. Hook, & P. Craddock (Eds.), *Metals and mines: Studies in archaeometallurgy* (pp. 211–218). London: Archetype.

Roberts, B., Thornton, C., & Pigott, V. (2009). Development of metallurgy in Eurasia. *Development, 83*(322), 1012–1022.

Robion-Brunner, C, Serneels, V and Perret, S. (2013). Variability in iron smelting practices: Confrontation of technical cultural and economic criteria to explain the metallurgical diversity in the Dogon Area (Mali). World of iron conference, Natural History Museum, London.

Robson, E. (2001). Society and technology in the Late Bronze Age: A guided tour of the cuneiform sources.

Rostoker, W., & Bronson, B. (1990). *Pre-industrial iron: Its technology and ethnology*. Philadelphia: Archeomaterials.

Rowlands, M., & Warnier, J. P. (1993). The magical production of iron in the Cameroon Grassfields. In T. Shaw, P. J. J. Sinclair, B. Andah, & A. Okpoko (Eds.), *The Archaeology of Africa: Food, metals and towns* (pp. 512–550). London: Routledge.

Scheel, B. (1989). *Egyptian metalworking and tools*. Oxford: Shire Publications.

Schmidt, P. R. (1996). *The culture and technology of African iron production*. Gainesville: University Press of Florida.

Schmidt, P. R. (1997). Iron technology in East Africa: Symbolism, science, and archaeology. Bloomington: Indiana University Press.

Schmidt, P. R., & Avery D. H. (1978). Complex iron smelting and pre-historic culture in Tanzania. *Science, 201,* 1085–1089.

Schmidt, P. R., & Avery, D. H. (1983). More evidence for an advanced prehistoric iron technology in Africa. *Journal of Field Archaeology, 10* (4), 421–434.

Schmidt, P. R., & Mapunda, B. B. (1997). Ideology and the archaeological record in Africa: Interpreting symbolism in iron smelting technology. *Journal of Anthropological Archaeology, 16*(1), 73–102.

Serneels, V. and Perret, S. 2003. Quantification of smithing activities based on the investigation of slag and other material remains. *Archaeometallurgy in Europe. Proceedings of the International Conference* (Milano, September 24-26, 2003). Milano: Associazione Italiana di Metallurgia, Vol. 1, pp. 469–478.

Severin, T., Rehren, T., & Schleicher, H. (2011). Early metal smelting in Aksum, Ethiopia: Copper or iron?. *European Journal of Mineralogy, 23*(6), 981–992.

Shinnie, P. L. (1985). Iron working at Meroe. In R. Haaland & P. L. Shinnie (Eds.), *African iron working: Ancient and traditional* (pp. 28–35). Oslo: Norwegian University Press.

Soper, R. (2002). Nyanga: Ancient fields, settlements and agricultural history in Zimbabwe. Nairobi: British Institute in Eastern Africa.

Stahl, A. B. (2014a). Africa in the World: (Re)centering African history through archaeology. *Journal of Anthropological Research, 70* (1), 5–33

Stahl, A. B. (2014b). Metal working and ritualization: Negotiating change through improvisational practice in Banda, Ghana, AD 1300-1650. In A. B. Stahl, *The materiality of everyday life. Archaeology papers of the American Anthropological Society*. Arlington: American Anthropological Society.

Summers, R. (1969). *Ancient mining in Rhodesia and adjacent areas*. Salisbury: Trustees of the National Museums of Rhodesia.

Thondhlana, T. (2012). Metalworkers and Smelting Precincts: Technological Reconstructions of Second Millennium Copper Production Around Phalaborwa, Northern Lowveld of South Africa. Unpublished Doctoral dissertation, University of College London.

Tylecote, F. F. (1980). Furnaces, crucibles and slags. In T. E. Wertime & J. D. Muhly (Eds.), *The coming of the age of iron* (pp. 183–228). New Haven: Yale Univ. Press.

Tamari, T. (1991). The development of caste systems in West Africa. *Journal of African History, 32* (2), 221–250.

Van der Merwe, N. J. (1980). The advent of iron in Africa. In T. E. Wertime & J. D. Muhly (Eds.), *The Coming of the Age of Iron*, (pp. 463–506). New Haven CT: Yale Univ. Press.

Van der Merwe, N. J., & Avery, D. H. (1987). Science and magic in African technology: traditional iron traditional iron smelting in Malawi. *Africa, 57*(2), 143–172.

Van Noten, F. (1985). Ancient and modern iron smelting in Central Africa: Zaire, Rwanda and Burundi. In R. Haaland & P. L. Shinnie, *African Iron working—ancient and traditional* (pp. 102–120). Oslo: Norwegian University Press.

Wagner, D. 2008. Science and civilisation in China. Vol. 5: Chemistry and chemical technology. Part 11: Ferrous metallurgy. Cambridge University Press. http://donwagner.dk/

Warnier, J. P., & Fowler, I. (1979). A nineteenth-century Ruhr in Central Africa. *Africa, 49*(4), 329–351.

Wobst, H. M. (1978). The archaeo-ethnology of hunter-gatherers or the tyranny of the ethnographic record in archaeology. *American Antiquity, 43*(2), 303–309.

Woodhouse, J. (1998). Iron in Africa: Metal from nowhere. In G. Connah (Ed.) *Transformations in Africa: Essays on Africa's later past* (pp. 160–185). London: Leicester University Pres.

Chapter 5
Socializing Metals

Introduction: Fabricating Metal into Cultural Products

The diversity in smelting practices observed for different peoples and periods in the history of different parts of Africa (Fig. 5.1), with their variably entangled and distinct technological traditions, characterizes processes of smithing—the next and culturally conditioned stage in the *chaîne opératorie* of metal working, to which this chapter is devoted. Sub-Saharan Africa differs remarkably from some of the Egyptian, Nubian and North African practices. In this region, as we have seen in the previous chapter, the process of smelting was a cultural intervention that transformed products of nature into culture. This act of transformation was symbolically associated with giving birth. The concept of 'bringing into life' was a fundamental one because, analogically speaking, the newly smelted metal (iron) passed through several additional stages of transformation (Herbert 1993). For example, metals were consolidated into ingots, or transformed into objects used in utilitarian, decorative and ceremonial domains (Cline 1937; Miller 2002; Reid and McLean 1995; Stayt 1931). In other contexts, ingots themselves were the final products used in religious and expressive spheres. Overall, the metal from furnaces and crucibles was traded sometimes as ingots but in other cases was smithed into objects. Egypt participated in the long-distance trade around the Mediterranean and also extending to Nubia, particularly after 3000 BC. Similarly, various sub-Saharan regions were networked locally and trans-regionally. Taken together, metals were situated within the nexus of society in that their use extended to utilitarian and nonutilitarian domains and were critical elements in the operations of economic, social, political, cultural and religious domains.

But how were the metals from Africa's diverse furnaces and cultures fabricated into disparate objects that addressed society's needs? This chapter is primarily aimed at addressing this question—it discusses the processes and techniques involved in African preindustrial metal smithing and fabrication. Metallographic and compositional work performed on a large corpus of objects and recovered from different parts of Africa illuminate the techniques of hot and cold working metal

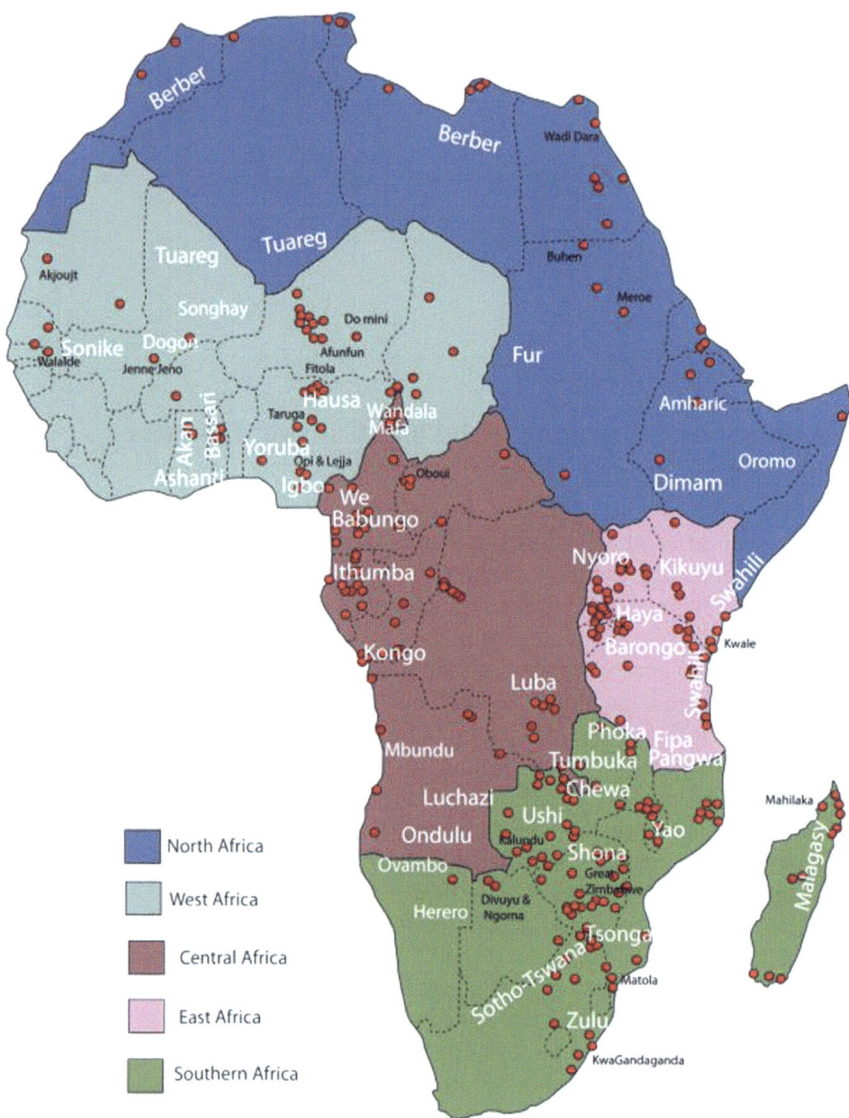

Fig. 5.1 Metal working groups and important sites discussed in the text

that were in use across the ages (Miller 1996, 2000; Scheel 1989). While hot working was practiced across the entire spectrum of metals and alloys that were manipulated preindustrially, techniques such as casting were specific to those that could be melted using available technology (Scott 1991).

In Egypt, pictorial paintings show the fabrication and casting of copper, gold, bronze and silver from Dynastic to Ptolemaic times (Scheel 1989). According to Miller (2002), techniques employed to fabricate metals in preindustrial sub-Saharan

Africa were fairly stable through time, although there were major instances of regional and temporal variation. Steel, the alloy of iron and carbon, was the hardest metal known to preindustrial sub-Saharan Africa (David et al. 1989). Iron and low-carbon steels were smithed into utilitarian and nonutilitarian objects. The use of finished objects was, however, not monolithic because so-called utilitarian items such as hoes were often used as currency in trade as well as in ritual and ceremonial settings (Dewey 1991). Other metals and alloys—copper, gold, bronze, brass and tin—were either smithed or casted into symbolic and decorative objects. Their colours appealed to various senses and valuations, resulting in a strong association with different genders, ancestors and power (Herbert 1993).

As with elsewhere in the Old World, preindustrial African metal smithing involved heating billets of metal in oxidizing environments (Crew 1991). This was followed by cycles of hot and cold hammering, which brought the metal to shape. The fabrication of copper, copper-based alloys and gold was designed to exploit not only their relatively low melting temperatures but also their physical properties such as ductility and malleability (Scott 1991). Routinely, these nonferrous metals were cast to produce a wide range of spectacular objects such as ingots, sheets, bangle blanks, beads and earrings (Miller 2002; Oddy 1984). Often, metals and alloys were drawn into wire or were hammered into thin sheets. In Southern Africa, these were wound around a vegetal core to produce spectacular decorative products (Miller 1996).

Over 7000 years of metal fabrication in Egypt and Nubia left residues and was often painted on tomb walls, which has remarkably preserved the evidence. In sub-Saharan Africa, over 2000 years of indigenous metal smithing has left ubiquitous fingerprints in the form of finished objects and tools (Garrard 1980; Kusimba and Killick 2003; Shaw 1970; Thondhlana and Martinón-Torres 2009). The contexts from which such traces of metallurgy were recovered powerfully demonstrate that as a technological solution, metal fabrication was precipitated by the need to achieve social, economic and political ends through the use of metals. This is testament to the important role played by metals in the preindustrial world. The integrated nature of metallurgy and society not only promoted or demoted the fortunes of smiths, but also enhanced the opulence enjoyed within varying gradients, by elites and commoners alike. In many sub-Saharan contexts, smiths were either feared or held in awe for their supernatural powers (McNaughton 1993), but in Egypt and Nubia, they formed a lowly class below that of learned officials and royalty (Scheel 1989), further underscoring the need to be attuned to variation in practice through time and across space.

Metal Fabrication in Egypt and Nubia

Because of their comparatively deep history of metallurgy—longer than anywhere else on the African continent–Egypt and Nubia offer interesting perspectives on the evolution of metal manipulation and fabrication. Scientific work suggests that the earliest objects of gold, native copper and meteoric iron were hammered and annealed (Craddock 2000; Rehren et al. 2013). With the advent of reductive smelting, metallic copper was sometimes hammered hot and cold but in other cases cast to produce objects, for temples, royalty, commoners and the military to mention a few (Ogden 2000). From the Old Kingdom (c. 3050BC) up to the end of Ptolemaic times (AD 385), a significant amount of pictorial depictions and associated texts in tombs demonstrates the techniques by which copper, tin bronze, gold, leaded tin bronze and silver were worked (see for example de Graris Davis 1943). During this period, metalworking was closely regulated by the state such that no independent specialists existed. In retrieving and returning stock, metalworkers and officials had to weigh opening and closing amounts and were always symbolically watched by Maat, the god of justice, who crowned the balance masts (Scheel 1989). Gold, silver and electrum served throughout Egyptian history as the basic materials for objects of royal use or for funerary and temple equipment (Aldred 1971), while copper, arsenic copper and bronze were used for the production of tools for daily use. As part of a *chaîne opératoire,* Egyptian and Nubian metalworkers melted large ingots or pieces of metal from furnaces or acquired through trade to refine, alloy, cast or split them up into smaller portions for further treatment by the smiths (Scheel 1989).

Exploiting the Behaviour and Properties of Metals: Melting, Casting and Plate Production

Scientific analysis of Egyptian artifacts has revealed that copper casting was practiced as early as the late Naqada II and Naqada III, between 3300 to 3000 BC (Scheel 1989). Copper tools and weapons were manufactured by hammering or open-mold casting. From the beginning of the Dynastic Period in Egypt around 3050 BC, metal fabrication techniques continued to be developed and improved, resulting in notable continuity and change. The metals (copper, tin, gold, silver, lead and their alloys) were melted in one or more crucibles, depending on the amount required (Scheel 1989). The pictorial depiction on the Sixth Dynasty tomb of the Vizier Mereruka at Saqqâra shows six metalworkers fanning through blowpipes into crucibles placed side by side. Although unverified, it is possible that fans of foliage were employed to provide a draught. In the Old and Middle Kingdoms, before the advent of artificial bellows, Egyptian metalworkers blew air into blow pipes consisting of reeds tipped with clay. Subsequently, skin bellows were used to provide the draught from the Middle Kingdom onwards. These were followed by dish bellows whose earliest depiction appears inside the Eighteenth Dynasty tomb of the priest Puyemre, the Second Prophet of the god Amun in Thebes (Fig. 5.2). The introduction of dish bellows enabled large quantities of metal to be melted for the casting of large metal objects as shown on a wall painting in the tomb of the Vizier Rekhmire at Thebes (Scheel 1989).

Although richly illustrated on tomb walls, finds of metal working sites are very rare. One ancient Egyptian foundry dating to Ptolemaic times was discovered in the Theban necropolis. The most important finds include numerous mud-brick hearths, which acted as receptacles for the burning charcoal onto which crucibles

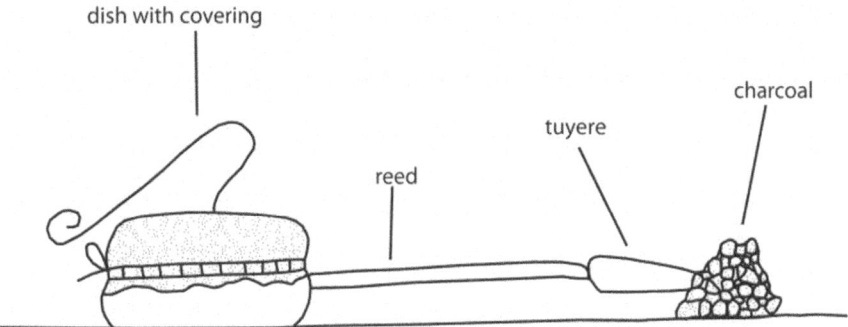

Fig. 5.2 Dish bellows connected to a clay nozzle with a reed (redrawn from Scheel 1989)

were placed (Scheel 1989). The foundry was designed for small-scale and mass production. Other finds include charcoal from acacia species, crucible sherds, broken tuyeres, limestone mold and the nozzles of dish bellows.

After being melted, refined and subsequently divided into portions, the cooled metal was passed to the smiths or blacksmiths for plate or sheet production. The metal was beaten on basalt, diorite or granite anvils, which were placed on a wooden block to absorb the hammering. Ancient Egyptian metalworkers used flat (for smoothing) and rounded hammer stones (for chasing). Evidence suggests that they practiced the technique of annealing as early as the Predynastic Period. Work-hardened metal was heated or annealed in a blowing or bellows-fanned fire to soften it to restore ductility. One pictorial in the tomb of Rekhmire depicts gold beaters placing thin gold plates on stone blocks and using hammer stones to repeatedly beat the gold leaf until the desired thinness was achieved (Scheel 1989). Silver and electrum also were worked to foil, while objects of less rare material were often gilded or silver-plated. Gold, silver and electrum foil or leaves were used to cover wooden furniture, statues, coffins and models of daily life manufactured for funerary equipment (de Graris Davis 1943). During the Roman Period, metalworkers practiced fire gilding with gold amalgam. Gold amalgam was applied to the base metal object to be gilded, and in the process, the mercury content of the amalgam vaporized, leaving gold attached to the surface of the metal objects (Scheel 1989).

From the Fourth Dynasty, ancient Egyptians joined together various components of objects using the techniques of soldering and riveting. It appears that Egyptian smiths applied different mixtures of gold, silver and copper to produce solders of different colours and melting points. The objects were polished on anvils using stones such as agate to smooth uneven patches on metal objects. Often, finished vessels were engraved with hieroglyphic text and other decorations. The engraver worked out the outline drawing using a hammer stone and chisels of different sizes.

Casting is yet another important metalworking process practiced by ancient Egyptians and Nubians. A diachronic study of casting shows that founders in Early Dynastic times poured molten copper or arsenic copper into preformed stone or clay molds to make simple tools and weapons. In the Old Kingdom, a more sophisticated form of casting using two-part molds of clay and steatite or serpentine allowed both faces of the object to be fashioned. Perhaps the most complicated method of casting is known as lost-wax or *cire perdue* casting, which enabled production of very detailed and complex shapes (Scheel 1989). This technique produced complicated jewelry, practical objects and statuettes for religious purposes. Initially, a beeswax model of the object to be cast was produced, and then coated with clay. The composite structure was heated in a charcoal fireplace to harden the clay and to melt the wax. In the process, the clay mold remained and retained all the details of the molten model. Egyptian founders then poured the molten metal into the clay mold. Upon cooling, the clay mold was broken and the cast object was cleaned and polished. Jewelry, amulets or other valuable items made of gold, silver and electrum were cast in this way. The technique of core casting was applied in casting large objects (Scheel 1989). A core of clay or sand was covered by a layer of malleable wax, shaped to form the mold of the object to be cast. The model, consisting of the core and the formed wax layer, was coated with clay. In order to secure it in position throughout the later casting process, the core had to be stabilized by pins or wire fixed to the outer cover of clay. The wax was melted away, leaving in the kiln the hardened clay mold with the fixed core. In casting, only the gap between the outer clay mold and the inner core had to be filled with bronze or other molten metal. Core casting was very common in the manufacture of larger objects during the New Kingdom and reached its peak in the Late Period (713–332BC). All these techniques were also exported to the Nubia (Emery 1963,1971)

Wirework and Jewelry

Gold, silver, and bronze wire was an important product of Egyptian metalwork. According to Scheel (1989), wire made its appearance during the First Dynasty and was used for mundane activities in the household, temples, palaces and in construction. Wire was produced through several methods. Initially, a metal piece was hammered out to sheet metal, which was then cut into thin strips. These strips were hammered out and cut again. Hammered wires show variations in diameter along the length, a faceted surface and a solid but noncircular cross section (Scheel 1989). This process of hammering was used to manufacture relatively thick wires. Another technique utilized in ancient Egypt is that of block-twisting, which appeared in New Kingdom times. The procedure involved hammering an ingot to create a square rod which was twisted about its major axis to form a solid wire with a screw thread of variable pitch. In the course of this twisting, the wire had to be annealed repeatedly to preserve its ductility. Spiraling could be eliminated by rolling the wire between two flat pieces of hard wood (Scheel 1989). In addition to hammering and block-twisting, strip-drawing and strip-twisting were probably mastered by Egyptian wire makers. A strip of metal foil was drawn through a number of holes of different diameters so that the strip curled in upon itself, forming a hollow tube. The strip could have been drawn through holes drilled through precious stones. In the final stage of wire production by this method, the wire could be drawn through holes of different diameters. Goldsmiths, silversmiths, and jewelers used wires of precious metal to decorate valuable objects or to manufacture pieces of jewelry such as anklets, aprons, armlets, belts, collars, chains, necklaces and pectorals.

Organization of Metalworking

Ancient Egyptian and Nubian metalworkers were not independent specialists; they were supervised by the state through a number of officials. Workshops were attached to temples, royal palaces or to the household of a high official. Because of their attached status, craftsmen in ancient Egypt were all dependent on their employers, who kept and allocated raw materials. Not surprisingly, metalworkers occupied a low position in the occupational specialization when compared to learned officials (Scheel 1989). This low social position contrasts with some areas of sub-Saharan Africa such as Central Africa where smiths were associated with royalty (de Maret 1985). Metalworkers were skilled in a range of metals and in relation to the physical properties of individual metals and alloys; they transferred skills from gold and copper to electrum, silver and bronze and eventually iron. Therefore, the copper worker was the gold worker who was also a bronze worker and later an iron worker (Scheel 1989).

Forging, Smithing and Casting in Sub-Saharan Africa: West and Central Africa

From their initial establishment in West and Central Africa, sometime in the radiocarbon black hole between 800 BC and 400BC, iron prevailed for utilitarian purposes, while copper was used for ornamental activities. Tin, bronze, gold and brass, introduced a millennium after iron and copper, were restricted in use to ornamental functions. As discussed above, this developmental trajectory differed from that of Egypt, Nubia, North Sahara, Ethiopia and Eritrea, which started with copper followed by bronze and iron. While metalworkers at Akjoujt in Mauretania started with copper, this was not followed by bronze, illustrating variation within the subcontinent. Unlike Egypt where a long record of literacy and tomb paintings provides insight into the fashioning of objects from metals, in West and other parts of sub-Saharan Africa, we have to rely on ethnography, travelers' reports and archaeological and technological studies for insight.

Two decades ago, Miller and Van der Merwe (1994) argued that traditional smithing and forging had received less attention in sub-Saharan technical literature than smelting, and the picture has not changed very much. While much is known from ethnographic sources, researchers should resist the present argument that ethnographically documented techniques represent unchanging continuities from earlier periods. Nevertheless, the ethnographic accounts of iron, copper and gold smithing are a good starting point for approaching variation in the archaeological past.

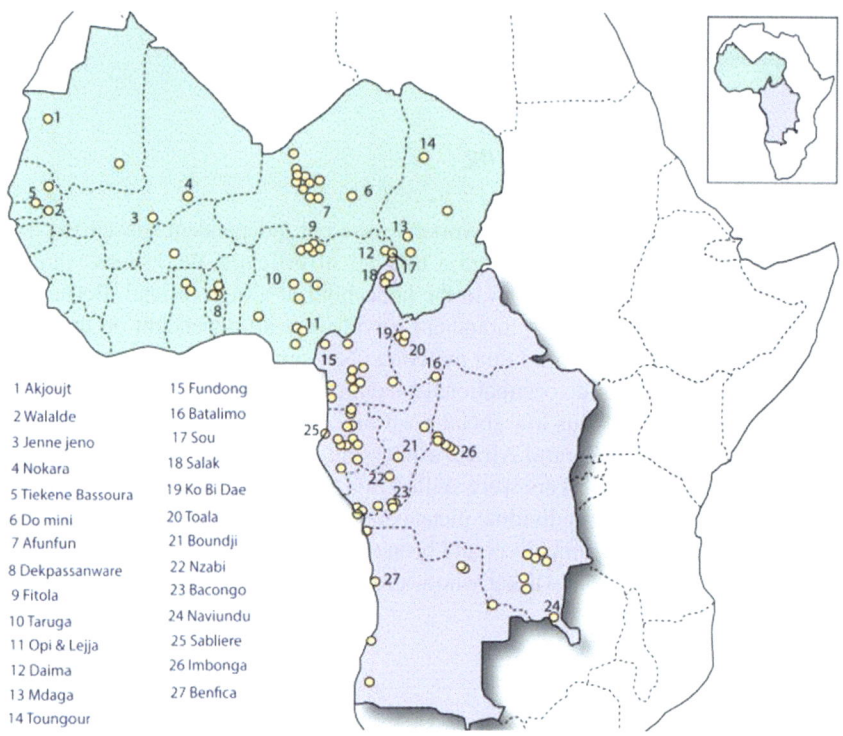

Fig. 5.3 Location of West and Central sites and groups discussed in the text

As in Egypt and Nubia and elsewhere in the world, West African smiths manipulated smelted, and melted metal in a number of ways: hammering (iron, gold and copper and its alloys), casting (copper, bronze, gold and brass) and wire drawing (copper, iron, gold, bronze and brass). In the area around Togo and Burkina Faso, the forges used from the sixteenth century AD onwards were very small and powered by concertina bellows (Cline 1937; Fig. 5.3). These forges contrasted with the very high natural-draught-powered smelting furnaces used in this area. The process of fabricating objects of iron started with heating blooms from furnaces on anvils to expel unwanted and occluded slag. The refined billet of metal was repeatedly hammered and annealed to gain usable metal. Today, smiths at places such as Bitchabe in Togo forge scrap iron using concertina bellows (Fig. 5.4).

Bellamy and Harbord (1904) described indigenous bloom forging and the smithing practice of the Yoruba of Oyo in early twentieth century Southwestern Nigeria. The forge was blown with a pair of valveless, circular, wooden bowl bellows covered with goat skin and pumped by centrally attached bamboo rods, feeding air to the open hearth through two wooden tuyeres. An iron hammer was used to beat the heated bloom, first on a large stone anvil and then on a smaller metal one to fashion the required tools. The indigenous metal was considered to be superior and more durable than imported English iron which contained sulfur from coal. In the north-

Fig. 5.4 Traditional iron forging workshop at Bitchabe, Togo, using concertina bellows. (Photograph Credit: Philip de Barros)

eastern village of Sukur, Northeastern Nigeria, smith's heated iron fragments in the forge for about half an hour (Sassoon 1964). The hot metal was beaten on a rock anvil with large hammer stones. The consolidated lump was then reheated and forged into the required tools. According to David and Sterner (1997), during the peak of the forging season, Sukur exported 60,000 hoes annually. In the nearby Mandara Mountains of Cameroon, the Mafa first sorted the products from the down draught furnaces into soft and cast iron. The cast iron was decarburized in crucibles to produce low-to high-carbon steels with typical martensitic microstructures. These were forged to produce iron bars which were traded over long distances. These smiths also produced shackles that were used in the slave trade (MacEachern 1993). Interestingly, while smelting in these areas took place in very large furnaces, forges were comparatively much smaller, yet produced a significant amount of finished products that fed into local, regional and in some cases long-distance trade. Therefore, this challenges assumptions that large-scale production of necessity equates with large-scale installations. Unlike smelting which could operate using natural draught, forging was invariably powered by bellows (Chirikure et al. 2009).

The available but sporadic evidence indicates that the technique of hammering was practiced in both the Early and Late Iron Ages of West and Central Africa. The smithing technology of the Early and Late Iron Age components at Dekpassanware was identical to ethnographically documented practice as revealed by finds of conical tuyeres, stone anvils, stone hammers and bloom crushing mortars (De Barros

2013, p. 17). Metallographic analyses of an Early Iron Age bracelet from a burial at Dekpassanware indicated that it was made of good-quality iron of variable carbon content. The microstructural evidence showed that the object cooled slowly and was therefore not quenched. The array of iron objects manufactured in West and Central Africa included hoes, iron bars, bangles, anklets, shackles, chains, spears, sheets and stout wire. Iron gongs, both single and double, were also made in some regions where they were associated with the power of kings (Vansina 1969).

The available evidence indicates that copper, bronze and gold were also hammered to produce a wide variety of objects in West and Central Africa including sheets, blanks, beads and wire (Cline 1937). More importantly, copper, gold and bronze and later brass were cast in open ceramic and steatite molds to produce simple shapes. The famous Katanga crosses of Central Africa dating from the mid- to late first millennium AD were cast using this method. By the late first millennium AD, the 'lost wax' process was used to cast copper and its alloys, with an early example supplied by bronzes at Igbo Ukwu in Nigeria dated to the nineth century AD (Shaw 1970). The famous Benin 'bronzes' and Akan gold weights were similarly produced through lost wax casting from the second half of the second millennium AD onwards.

During the lost wax process, the desired object was either molded using wax or wax was used to cover a clay model mold. Subsequently, support rods of wax were attached to the exterior of the sculpture to create casting sprues on the wall of the mold. These enabled the cast metal to flow without trapping air pockets. In similar fashion to the practice of ancient Egyptian metal workers described above, the entire structure was dipped in liquid clay mixed with finely ground ash and, when dry, coated in a thick clay jacket. The dried mold was then fired to remove the wax/latex, after which molten metal was poured into the resultant void. In some cases, the crucible used to melt the metal was physically attached to the mold (Bisson 2000). On cooling, the clay exterior was shattered, thereby exposing the sculpture. The object then had the casting sprues removed and was subsequently cleaned and polished. This technique was finely executed to produce the famous Igbo Ukwu copper and bronze objects in Nigeria dating to between the nineth and thirteenth centuries AD (Shaw 1970; Fig. 5.5). The Igbo Ukwu bronzes represent a masterpiece in terms of metalworking, are easily the most technically skilled castings in all of sub-Saharan Africa, and are among the finest art castings ever done anywhere in the world at any time (Killick pers comm 2014). Thurstan Shaw's excavations unearthed more than three hundred elaborately cast objects including the renowned roped pot (Fig. 5.5), immaculate bronze altar stands, spectacular bronze bowls, elaborate bracelets, copper snakes, a bronze snail, bronze elephant, bracelets and much more. Over 80 objects from Igbo Ukwu were analysed, and those produced using lost wax were leaded bronzes, while the hammered and chased ones were made of pure copper (Shaw 1970). Other objects include very fine copper and bronze wire work. Igbo Ukwu also yielded significant amounts of hammered iron objects, pieces of ivory and thousands of trade beads, indicating that it represents high-status burials. Initially, it was believed that the Igbo Ukwu bronzes were not local in origin. However, compositional and provenance studies indicated that the copper was obtained from

Fig. 5.5 Leaded tin bronze Igbo Isiah roped vessel produced using the lost wax method (height: 32.2 cm). (Photograph Credit: Estate of Thurstan Shaw with permission of Dr Pamela Smith)

the Benue River valley, while the tin came from the Jos Plateau. Lost wax technique continued to be practiced in what is today Nigeria well into the nineteenth century and resulted in the equally outstanding bronze and brass objects from Ife and Benin (Chikwendu et al. 1989).

According to Garrard (2011), the stories of abundant gold in West Africa, particularly from Islamic writers, has not matched the recovery of gold from archaeological sites. Amongst the few gold objects recovered include a very fine pectoral from Rao Senegal (Fig 5.6) and a seventh to nineth century earing from Jenne Jeni in the Inland Niger Delta of Mali (Fig 5.7). Reports by Al Bakri in the eleventh century indicate that gold was exported from West Africa to North Africa via the town of Tewdagoust (Nixon et al. 2011). The scale of West African production was revealed when Mansa Musa the king of ancient Mali (AD1200–1400) made pilgrimage to Mecca with significant quantities of gold, which led to a dramatic fall of gold prices in Cairo (Levtzion 1973) and ultimately whetting the European appetite for gaining direct access to African gold. While historical focus has tended to be on the role

Fig. 5.6 Pectoral from the tumulus of Rao Senegal dating between seventeenth and eighteenth centuries. (Source: Gold of Africa Museum, Cape Town)

Fig. 5.7 Photograph showing earliest gold earing from Jenne Jeno, Inland Niger Delta, Mali. (Source: Rod McInstosh)

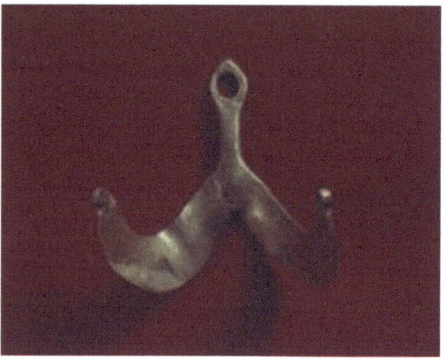

of gold as an export commodity, West African gold smiths produced technically sophisticated objects of considerable beauty. One of the most spectacular is a gold pectoral excavated by archaeologists at the tumulus of Rao in contexts dating between the seventeenth and eighteenth centuries (Fig. 5.6, Garrard 2011). In the area around modern Ghana, the Akan people produced spectacular gold objects, which like brass objects worked by the same group, were hammered into thin sheet or cast (Garrard 2011, p. 132).

Just as the skilled metal workers were attached to temples/political authorities in Egypt, in some West African contexts, prestige metals such as gold and brass were

Fig. 5.8 Akan crocodile and lizard sword ornaments (size 21.5 cm, see Garrard 2011, p. 234) (Source: Gold of Africa Museum, Cape Town)

intimately bound up in the institutions of state (Garrard 1980, p. 38). For example, Akan goldsmiths produced gold objects such as sword ornaments (Fig. 5.8), which were used in stately functions together with the Golden Stool (Garrard 1980, p. 63). Interestingly, the gold smiths and other skilled metallurgists practiced as attached specialists (Garrard 1980, p. 47).

East and Southern Africa with Occasional References to Sudan and Ethiopia

Turning to fabrication processes in East and Southern Africa (Fig. 5.9), a short account of smithing in Sudan recorded by Crawhill (1933) documented important features of forging.

The fragments of iron blooms from the smelting furnace were heated in open clay crucibles until they were soft enough to be hammered together between stone anvils and then shaped further using a handle-less iron hammer. The smiths used four valve-less hand-operated bellows whose cylinders were made of clay and covered with goat skin diaphragms. In Ethiopia, Dimi forges of the twentieth century consisted of a walled hearth driven by two drum bellows operated by hand that blew air into a single tuyere (Todd 1985). A blunt iron hammer and stones were used to fashion the desired implement by hammering the heated bloomery iron on a specially selected stone anvil. The final product was not quenched or tempered. Broken tools were also repaired at the forge. Brown (1995) studied ethnographic iron working among the Kikuyu in Kenya in the twentieth century. The Kikuyu forged iron in small forges powered by bellows. They used stone anvils and hammer stones. They also quenched the finished objects in water.

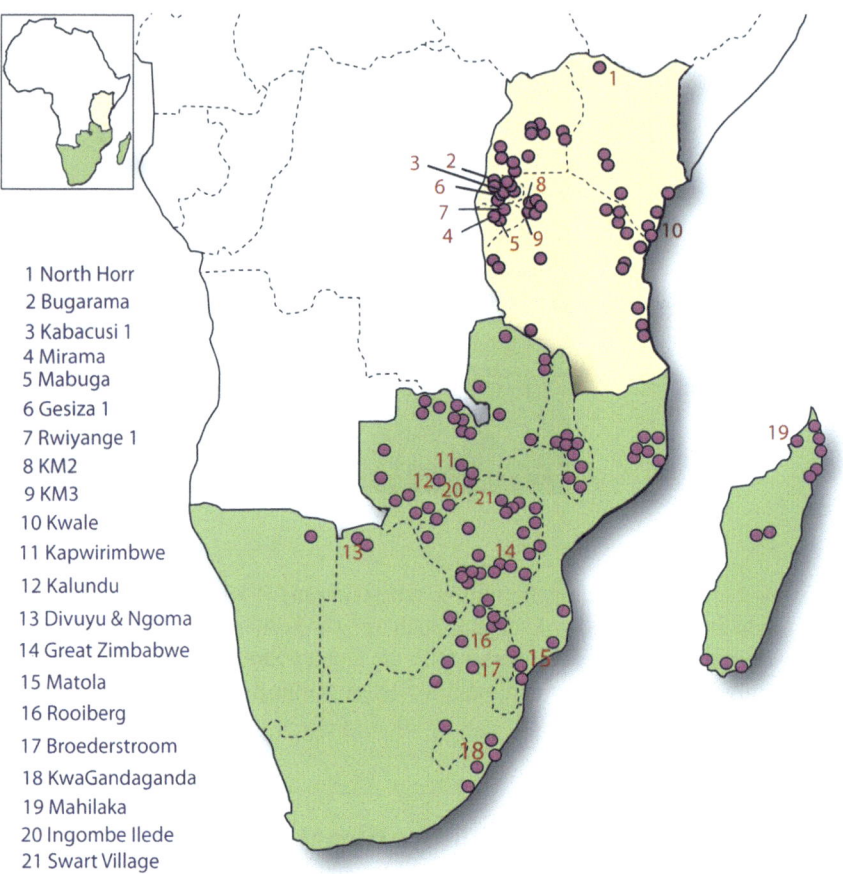

1 North Horr
2 Bugarama
3 Kabacusi 1
4 Mirama
5 Mabuga
6 Gesiza 1
7 Rwiyange 1
8 KM2
9 KM3
10 Kwale
11 Kapwirimbwe
12 Kalundu
13 Divuyu & Ngoma
14 Great Zimbabwe
15 Matola
16 Rooiberg
17 Broederstroom
18 KwaGandaganda
19 Mahilaka
20 Ingombe Ilede
21 Swart Village

Fig. 5.9 Location of smithing sites in East and Southern Africa

One of the best ethnographic descriptions of preindustrial iron smithing in Africa is provided by Emil Holub, a Czech medical practitioner who journeyed from Cape Town to the Barotseland plain of Zambia in the late 1870s. While in Zambia, he recorded the activities of a Mambari iron smith, paying meticulous attention to the pumping of the bellows and the tools used in the process. Holub (1881) made impressive illustrations not only of these bellows but also of the other tools used by the smith. These include hammers, chisels, tongs, anvils and some finished objects (Figs. 5.10 and 5.11).

South of the Zambezi, Njanja iron forging was discussed by a number of researchers. After a smelting re-enactment carried out by Headman Ranga in the 1950s, a bloom from the furnaces was consolidated in a forge, powered by goat skin bellows (Fig. 5.12). The bloom was repeatedly hammered on an anvil until the metal was usable. Iron hoes were made by forge welding pieces of metal together. Forge welding resulted in the incorporation of surface scale into the weld lines

East and Southern Africa with Occasional References to Sudan and Ethiopia

Fig. 5.10 Illustration of a Mambari smith by Holub (1881) showing bellows leading to tuyeres and a very small forge. Note also the tongs illustrated

Fig. 5.11 Illustration of pot bellows used for smithing by Mambari smiths (Holub 1881)

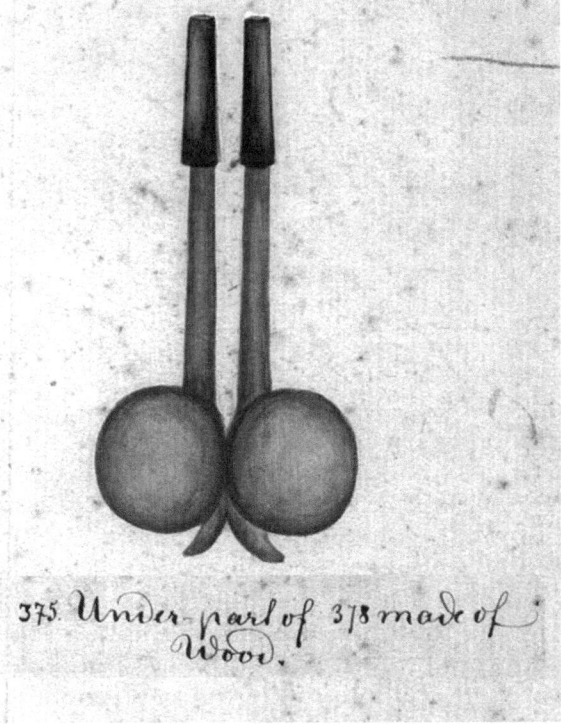

Fig. 5.12 Goat skin bellows similar to the ones used by Njanja. Approximate size 80 cm long. Natural History Museum, Bulawayo, Zimbabwe (Photograph Credit: Author)

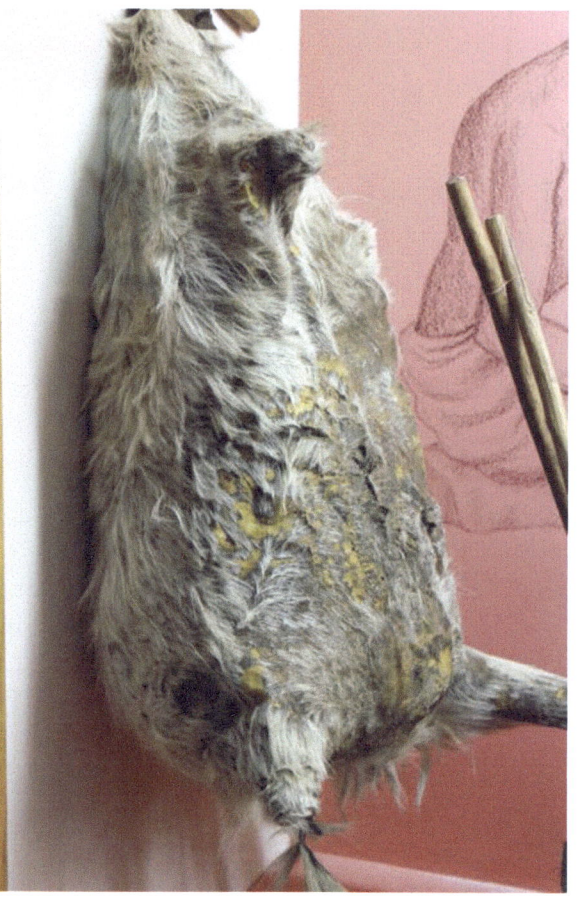

which created internal voids in the objects. This enables archaeometallurgists to identify objects manufactured using preindustrial techniques. Furthermore, bloomery iron consisted of slag stringers that are microscopically detectable. Objects such as axes were quenched in water, which improved the strength of the metal with some carbon in it. This may be why objects from Njanja smiths were highly regarded in areas with their own smiths (Chirikure 2006; Mackenzie 1975). Interestingly, Njanja scale of production is described as semi-industrial and was often characterized by the presence of itinerant smiths who smithed iron in different areas and in the process amassed wealth. Igbo smiths in Nigeria were similarly itinerating.

As with iron forging, copper and gold smithings also feature prominently in ethnographic and historical accounts of Southern Africa. In most of Southern Zambezia, copper was melted and cast in ingot molds that produced X-shaped copper ingots similar to those produced in Central Africa (Fig. 5.13).

The most vivid ethnographic descriptions of copper working are those of the Venda and Ba-Phalaborwa people, historically associated with parts of Southern

Fig. 5.13 X-shaped copper ingots in use in much of Southern Zambezia in the Iron Age, particularly after AD1000 (Photograph Credit Author)

Fig. 5.14 Musuku ingot housed at Iziko Museums, Cape Town (Photograph Credit: Author)

Zimbabwe and parts of Northern South Africa. The copper from furnaces was melted in ceramic crucibles that resemble domestic pots (Stayt 1931; Mamadi 1940). Once refined, the metal was either cast to produce small items such as bangle blanks or hammered to produce thin sheet. In some cases, the molten metal was cast to produce the iconic *musuku* and *lerale* ingots (Stayt 1931; Miller et al. 2001). *Lerale* ingots were made by pressing a 1-or 2-cm-diameter stick lengthwise into the soil. At one end of the mold, a small hollow was carved to create a small head, sometimes with short arms protruding from it. Molten metal was then poured, which on solidification produced *lerale* ingots (Miller et al. 2001). In some cases, the end of a larger stick was thrust into the ground, producing an oblong hole with straight sides and a flat bottom. The ends of smaller sticks were pressed into the bottom of this hole to form a pattern of smaller holes, usually parallel lines (Stayt 1931). When filled to overflowing with copper, this mold produced a *musuku* ingot, a short cylinder with studs on one end and an irregular flange where the copper had spilled out of the top of the mold (Mamadi 1940 Fig. 5.14). These objects were used in ritual and ceremonial settings associated with political authority and healing.

Ellert (1993) studied traditional gold production on the Zimbabwe plateau and observed that gold dust and nuggets were melted in ceramic crucibles resembling domestic pottery. It seems that very few specialized crucibles were used for gold working. By contrast to West African examples such as Asante where gold was a prerogative of state (Garrard 1980, pp. 135–136), gold production was not associated with any centralized authority for villagers could freely scour for ore and melt it for a variety of purposes (Mudenge 1974).

Archaeologically, a number of researchers have studied the microstructures and chemistry of finished metal objects in Eastern and Southern Africa. In particular, Kusimba et al. (1994) studied 180 iron artifacts from Swahili sites on the East African coast. While the techniques were generally similar to those practiced by other African communities, the late first millennium AD Swahili inhabitants of Galu and Ungwana worked cast iron which was decarburised to the forgeable low-carbon steels presumably in crucibles or oxidising hearths (Kusimba and Killick 2003, p. 180). Cold forging and annealing were also practiced together with pressure welding or joining two hot pieces of metal together. It has also been surmised that the Swahili at Galu may have produced crucible steel, which was used for more mundane tools such as nails (Kusimba and Killick 2003). While the Swahili's ability to produce high-carbon steel has never been questioned, it is possible that the crucible steel may have been imported (Killick 2009).

In Africa south of the Zambezi, Miller (1996; 2002) carried detailed metallographic and compositional analyses of iron and copper objects from Early Iron Age (AD200–1000) and iron, copper, bronze and gold objects from Late Iron Age sites (AD1000–1850). Most of the Early Iron Age objects came from Nqoma and Divuyu in Botswana, while analysed Late Iron Age ones had a larger geographical spread. The inventory of iron objects made in the 2000-year history of fabrication in Southern Africa include needles, awls, bracelets, bangles, spears, axes, arrowheads and hoes among others. Miller's detailed work revealed that small ferrous items were often fully heated or annealed as testified by the presence of spheroids in their microstructures, which normally result from cumulative overheating (Miller 1996). Archaeological and metallographic work has shown that the basic fabrication technology used for working indigenous bloomery iron, copper and gold did not change sharply over time, despite meaningful changes in the types of objects made (Miller 2002). While small objects prevailed at first-millennium AD sites, larger ones became ubiquitous in the second millennium, at sites such as Bosutswe, Mapungubwe, Khami and others. Presumably, during the first millennium AD, iron was still a rare metal, which was consistently recycled, and because it was widely available in the second millennium AD objects of various sizes could be found (Miller 1996).

Sometimes, softened (annealed) iron was forged into thin sheets and cut into long strips to make helical wound bangles, often using a flexible fiber core as well as clips and wrapped beads. Stout iron wire was made by hammering and rotating a thin rod (Miller 1996). Metallographic and compositional analyses of objects from Caton-Thompson's excavations at Great Zimbabwe by Stanley (1931) identified bronze, brass and leaded gun metal, which were all fabricated using indigenous techniques. Bronze and brass were certainly imports, and if the early twelfth century date is to be believed, imported metal was coming into the interior from the coast much earlier than some researchers are prepared to accept. This is hardly surprising, for bronze appeared on the East African coast at AD 700 (LaViolette 2008), and glass beads and other imports were coming into the interior from the eighth century AD, if not earlier.

Miller's (2002) study demonstrated that the techniques invested in fabricating iron were similar to those invested in copper (AD900–1850), bronze, and gold working (AD1000–1850). Routine copper, bronze and gold fabrication involved the alternate heating and hammering of ingots into shape. Often the finished objects were left to cool slowly, but sometimes hammered sheets of copper, bronze and gold were cut into thin strips, which were wound around vegetal cores to make bangles. In the Early and Late Iron Ages, copper beads were made by cutting short lengths of metal strip or thin rods with a chisel and bending them around with the cut levels on the inside to create a relatively smooth join without welding or soldering (Miller 2002; Thondhlana and Martinón-Torres 2009). After the second millennium AD, gold was also hammered into foil, which was attached to wooden poles and objects as shown by the examples from the Valley Enclosures at Great Zimbabwe (Chipunza pers comm) and Mapungubwe (Oddy 1984; Miller 2002).

Copper and gold were cast to produce preformed shapes. Molds and crucibles for working gold and copper have been found in Southern Africa. Crucibles used in most of Southern and Central Africa basically resembled normal pottery (Bisson 2000). Thondhlana (2013) excavated such crucibles dating to the Kgopolwe (early second millennium AD) levels at Shankare in Phalaborwa. Miller and Hall (2008) also found such crucibles at Rooikrans dating from the sixteenth century AD onwards. However, often there were also specialized crucibles such as the one from Phalaborwa, which has a pourer (Thondhlana 2013). There are also examples of sandstone crucibles, which were used for melting brass in the sixteenth and seventieth centuries. However, few studies of crucibles have been conducted in Southern Africa. Archaeologists have also recovered ingot molds (Fig. 5.15) below which is on display at the Natural History Museum in Zimbabwe. X-shaped ingots were

Fig. 5.15 Half of a cross-shaped copper ingot mold (carved from steatite?) used to produce ingots in 5.15 recovered from Zimbabwe, on display at the Natural History Museum, Bulawayo, Zimbabwe (Photograph: Author)

Fig. 5.16 Second-millennium AD soapstone ingot mold with a gold pellet insert, Natural History Museum, Bulawayo (Photograph credit: Author)

found at Great Zimbabwe indicating that the mold may date to any period in the second millennium AD.

Gold dust was melted in ceramic crucibles resembling domestic pottery and was often cast in soapstone molds to produce spherical beads (Fig. 5.16).

Of all the techniques used for copper, bronze and gold working, wire drawing seems to have been the most challenging to execute. It was carried out using plates some of which were oblong in cross section. According to Bisson (2000), wire draw plates had circular holes ranging between 3.5 cm and 1 mm in diameter. During the process of wire drawing, the draw plate was fixed to a stand. A heated copper rod, with its end specifically tapered for the process, was pulled through the draw plate using a pair of tongs. The resultant wire was pulled consecutively through a graded series of holes in the iron plate until the desired diameter was achieved (Brown (1995). Usually, the last grade was a millimeter thick, and this extremely fine gauge copper wire was an important constituent of spirally wound bangles, which are ubiquitous at Southern African archaeological sites. These are often found in large quantities in burials of high-status individuals (Fig. 5.17) from capitals such as Danangombe (AD1680–1850) and Mapungubwe.

Ingombe Ilede, a mid-second-millennium AD ossuary on the banks of the Zambezi River in Zambia (Fig. 5.1) has yielded some of the best evidence of copper

Fig. 5.17 Femur and bangles from a high-status individual excavated from the dry-stone-walled Zimbabwe culture site of Danangombe (AD1680–1850), Central Zimbabwe. The excavators retrieved this bone with those high numbers of copper or copper alloy bangles, which are quite numerous. It is on display in the Zimbabwe Museum of Human Sciences. Photograph: Author

working in Africa. Discovered in the 1960s, the site consists of various burial chambers. The accompanying grave goods speak of specialized copper workers who were buried with a complete set of wire drawing equipment, unused X-shaped copper ingots and an amazing copper bar that represented the first stages of the wire drawing process (Phillipson and Fagan 1969). To crown it all, finished copper wire was also part of the funerary goods.

As in West and Central Africa, iron was used to make utilitarian objects such as hoes, axes, hammers and among arrow heads while initially copper, but later brass, bronze and gold were used to make decorative and ornamental objects such as bracelets, beads, earrings, symbolic hoes such as the bronze ones from Khami and Great Zimbabwe, and thin sheets/foil. However, this was not a static dichotomy for iron was also used to make bangles, anklets and ceremonial objects such as gongs, axes and spears.

One striking difference between metalworking in Southern and West Africa is that while endogamous occupational groups or castes have deep roots in areas such as Mali (Robion-Brunner et al. 2013), such castes are not known in Southern Africa. The status of smiths was variable such that in some cases they occupied a high status and even founded chiefdoms as discussed by Mackenzie (1975) in the case of the seventeenth-century Njanja in Central Zimbabwe. Maggs (1992) discusses the example of iron making in the Zulu state where the Amalala were reduced to a class of attached specialists that produced metal for Shaka's army. However, even in this case, the Amalala did not transform into a caste.

This points to variability in the social status of metal workers across Africa and through time. Smiths and potters in Ethiopia and Sudan belong to the same caste regarded as lowly at the same time as feared (Haaland 2004b). Intermarriage between castes and non-caste members was forbidden. This case is analogous to that documented in West Africa by Tamari (1991) and others such as McNaughton (1993). In a different context, ancient Egyptian and Nubian metalworkers, as we have seen, occupied a low class, but this was not inherited. Anybody who worked hard and became learned could be an official, while those who failed became artisans (Scheel 1989). These differences in the organization of metalworking at different points of African history warn against the dangers of extrapolating observations from one area and time period into another. Generalizations must be finely calibrated in relation to context-specific situations; otherwise, they create reconstructions perfectly sensible to our minds but tangential with what may have happened in the past.

Metals in Society: The Anthropology of Smithing and Metal Objects

As Mauss (1954) has submitted, technology is a complete social and cultural phenomenon. As an important element of the *chaîne opératoire* of metal production and use, smithing, like smelting and mining, was also associated with cultural ideas of its own but which drew from the general symbolic load as articulated in broader society. Gods such as Maat played an important role in Egyptian and Nubian metalworking just as they did in life from Dynastic to Ptolemaic times. The metals themselves also had values associated with them from the valued gold, silver and electrum used in palaces, temples and tombs to iron which when established was used for day to day objects. The centralized organization of metal fabrication enabled through effective record keeping ensured that these states controlled metal working in ways that probably did not characterize empires such as ancient Ghana and Mali where gold producers appear to have operated independently and only remitted tribute to the capital (Levtzion 1973). This was probably also the case in relation to Mapungubwe, Great Zimbabwe, and polities that emerged afterwards. Here, despite centralized organization, the indication and absence of large-scale debris from capitals is that production took place in the hinterland. Lack of an effective means of record keeping may have comprised central control of production. As such, Mudenge (1974) argues within the context of the Rozvi State (AD1680–1850) that commoners produced and traded metal, only submitting taxes to the courts. The outcome of this situation was that some individuals became very wealthy with the result that metalworking was highly respectable in this area.

In terms of the human procreational paradigm, the reduced iron metal was metaphorically seen as a child in sub-Saharan Africa (Herbert 1993). Although smelting was sometimes practiced away from residents because of taboos and other factors, smithing took place in the village (Schmidt 2009). The smithing of objects took place in full view and was not shrouded in secrecy or mystery (Cline 1937). A variety of objects were fashioned from metal and these assumed a persona, for they played a central role in solving societal problems as well as in negotiating and structuring personal relations.

Metal objects gained value depending on the context in which they were used. For example, the iron hoe occupied an important role in Shona fertility and associated beliefs. A hoe or *badza* made it possible for the Shona to cultivate the land to harvest food, so too was it important for weeding the fields and digging the earth, placing it at the center of societal renewal and growth (Bent 1896; Bourdillon 1976). Not surprisingly, the iron hoe fundamentally structured human relations, particularly those related with fertility and growth. For some, iron hoes were an important medium in marriage transactions, as for example among the Shona for whom *roora* or *lobola* (bride wealth) is often referred to as *badza* or the hoe (Childs 1991). During marriage negotiations, a father whose daughter was about to be married demanded the symbolic iron hoe to ensure continued productivity of his household. The iron hoe too was an important a symbol of divorce. If a husband wanted to divorce his wife, he gave her a worn out hoe to take to her parents, a concept known as *gupuro* in Shona (Bourdillon 1976). In between these human relations, a hoe was a basic tool for achieving a variety of tasks in society, but because of these utilitarian and symbolic roles, it served a nodal role in a web of social relations. Indeed, the iron bundles known as *bikie* in Southern Cameroon also played a similar role in this part of the world (Guyer 2012).

A fundamental way in which metal is socialized is through giving names, often ones associated with the tasks and not the metal. This socialization is hardly surprising because metal symbolically represented nature that had been tamed through reduction. Once smithed into artifacts, such objects acquired names and meanings which had little or nothing to do with metal. A Shona copper or iron needle was known as *tsono* and it was used for sewing (*kusona*). A Shona iron axe was not referred to as an iron axe, rather it was known as an axe or *demo*. This process of socializing material culture equally applies to other African communities and those beyond (see Appandurai 1986). The forging of objects and their use socialized metal artifacts, placing them alongside other material culture whose role and function was more dependent on context than material. As such, the objects are fundamentally human and it is human to make, use, personify and discard the objects.

Of course, the role of metal objects far extends beyond the ideas summarized here. Metal became the pivot on which trade and exchange, interaction, culture contact, and even warfare were anchored. This broader theme is explored in detail in Chap. 6.

Conclusion

Building on the history of metalworking in Africa's multiple regions, copper in parts of West, Central, East and Southern Africa was largely reserved for ornamentation, or in some cases sculpture, while agricultural tools and weapons were always made of iron. In this respect, sub-Saharan Africa was very different from Eurasia in the Bronze Age, and even from Egypt and North Africa, where bronze weapons were widely used. Even after the introduction of gold and bronze, these too were restricted to the ornamental and ceremonial domains. In fact, it appears as if gold, with few exceptions (e.g., Asante; Garrard 1980), did not appeal to commoners in sub-Saharan Africa who preferred the red colour and tonality of copper when compared to gold (Herbert 1993). As a result, control of gold production was not one of the ways in which elites maintained their power. Rather, they controlled the land and its fertility through a link to their ancestors (Mudenge 1974). This way copper, iron and gold working together with agricultural success were linked to ancestors who bestowed this through their descendants in power. Such a situation also differs with Egypt and North Africa where rulers had to account for every bit of metal and reserving for their exclusive use the highly valued gold, silver and electrum.

Linked to the large-scale iron production in West and Central Africa as well as at Meroe in the Sudan is the semi-industrial production of iron billets at (1) Sukur where iron was traded north into the Lake Chad Basin (David and Sterner 1997); (2) Bassar in Togo for sale to Hausa caravans (de Barros 2013); (3) Meroe for Ptolemaic and Roman Egypt (Shinnie 1985); (4) the Cameroon grasslands (Warnier and Fowler (1979); and (5) among the Dogon in present-day Mali (Robion-Brunner et al. 2013). It is possible that high levels of production in West and Central Africa were a response to the slave trade (MacEachern 1993), but it was also linked with demographic increases, something which was not of consequence in Southern Africa where the scale of production was hardly comparable.

Finally, there are substantial intersections between techniques employed in various regions, just as there were major differences. For example, the lost wax technique, which represents the pinnacle of copper and copper alloy casting in Egypt and West Africa, was unknown in Southern and Eastern Africa. The techniques invested in hot or cold working were similar across the metals and alloys used in African civilizations. Differentials in the physical properties of the individual metal motivated for the casting of copper, tin and gold. These differences were stimulated by local innovations as well as by broader regional interactions about which we continue to learn through archaeological sources (Mitchell 2005).

References

Aldred, C. (1971). *Jewels of the pharaohs: Egyptian jewelry of the dynastic period*. New York: Praeger.
Appadurai, A. (Ed.) (1986). *The social life of things: Commodities in cultural perspective*. Cambridge: Cambridge University Press. (Archetype).
Bellamy, C. V., & Harbord, F. W. (1904). A West African smelting house. *Journal of the Iron and Steel Institute, 66,* 99–126.
Bent, J. T. (1896). *The ruined cities of Mashonaland: Being a record of excavation and exploration in 1891*. London: Longmans, Green, and Co.
Bisson, M. (2000). Precolonial copper metallurgy: Sociopolitical context. In M. Bisson, P. de Barros, T. C. Childs, & A. F. C. Holl (Eds.), *Ancient African metallurgy: The sociocultural context*. (pp. 83–146). Walnut Creek: AltaMira.
Bourdillon, M. F. (1976). *The Shona people: An ethnography of the contemporary Shona with special reference to their religion*. Gweru: Mambo Press.
Brown, J. (1995). *Traditional metalworking in Kenya*. Oxford: Oxbow.
Chikwendu, V. E., Craddock, P. T., Farquhar, R. M., Shaw, T., & Umeji, A. C. (1989). Nigerian sources of copper, lead and tin for the Igbo-Ukwu bronzes. *Archaeometry, 31*(1), 27–36.
Childs S. T. (1991). Style, technology and iron smelting furnaces in Bantu-speaking Africa. *Journal of Anthropological Archaeology, 10*(4), 332–359.
Chirikure, S. (2006). New light on Njanja iron working: Towards a systematic encounter between ethnohistory and archaeometallurgy. *South African Archaeological Bulletin, 61,* 142–151.
Chirikure, S., Burrett, R., & Heimann, R. B. (2009). Beyond furnaces and slags: A review study of bellows and their role in indigenous African metallurgical processes. *Azania: Archaeological Research in Africa, 44*(2), 195–215.
Cline, W. B. (1937). *Mining and metallurgy in Negro Africa (No. 5)*. Menasha: George Banta.
Craddock, P. T. (2000). From hearth to furnace: Evidences for the earliest metal smelting technologies in the Eastern Mediterranean. *Paléorient, 26*(2), 151–165.
Crawhall, T. C. (1933). Iron working in the Sudan. *Man, 48,* 41–43.
Crew, P. (1991). The experimental production of prehistoric bar iron. *Historical Metallurgy, 25*(1), 21–36.
David, N., & Sterner, J. (1997). Water and iron: Phases in the history of Sukur. In H. Jungraithmayr, D. Barreteau, & U. Seibert (Eds.), *L'homme et l'eau dan le basin du lac Tchad*. (pp. 255–270). Actes du Colloque Mega-Tchad. Paris: Colloques et Seminaires, Editions ORSTOM.
David, N., Heimann, R., Killick, D., & Wayman, M. (1989). Between bloomery and blast furnace: Mafa iron-smelting technology in North Cameroon. *African Archaeological Review, 7*(1), 183–208.
de Barros, P. (2013) A comparison of early and later Iron Age societies in the Bassar region of Togo. In J. Humpris & Th. Rehren (Eds.), *The world of iron* (pp. 10–21). London: Archetype.
de Garis Davies, N. (1943). *The Tomb of Rekh-mi-Re͑ at Thebes* (Vol. 2). New York: Metropolitan Museum of Art.
de Maret, P. (1985). The smith's myth and the origin of leadership in Central Africa. In R. Haaland & P. Shinnie (Eds.), *African iron working—ancient and traditional* (pp. 73-87). Oslo: Norwegian University Press

Dewey, W. J. (1991). *Weapons for the ancestors*. Des Moines: University of Iowa, Department of Art History.
Ellert, H. (1993). *Rivers of gold*. Gweru: Mambo Press.
Emery, W. B. (1963). Egypt exploration society preliminary report on the excavations at Buhen, 1962. *Kush, 11*, 116–120.
Emery, W. B. (1971). Preliminary report on the excavations at North Saqqâra, 1969–70. *The Journal of Egyptian Archaeology, 57*, 3–13.
Garrard, T. (1980). *Akan weights and the gold trade*. London: Longman.
Garrard, T. F. (2011) (1989). *African gold: Jewellery and ornaments from Ghana, Côte d'Ivoire, Mali and Senegal in the collection of the Barbier-Mueller Museum*. Munich: Prestel.
Guyer, J. I. (2012). Soft currencies, cash economies, new monies: Past and present. *Proceedings of the National Academy of Sciences, 109*(7), 2214–2221.
Haaland, R. (2004b). Iron smelting—a vanishing tradition: Ethnographic study of this craft in South-West Ethiopia. *Journal of African Archaeology, 2*(1), 65–79.
Herbert, E. W. (1993). Iron, gender, and power: Rituals of transformation in African societies. *History, 32*(2), 221–50.
Holub, E. (1881). *Seven years in South Africa: Travels, researches, and hunting adventures, between the diamond-fields and the Zambesi (1872–79)* (Vol. 2). London: S. Low, Marston, Searle & Rivington.
Killick, D. J. (2009). Agency, dependency and long-distance trade: East Africa and the Islamic World, ca. 700–1500 C.E. In S. Falconer & C. Redman (Eds.), *Polities and power: Archaeological perspectives on the landscapes of early states* (pp. 179–207). Tucson: University of Arizona Press.
Kusimba, C. M., & Killick, D. (2003). Iron working on the Swahili Coast of Kenya. In C. M. Kusimba & S. B. Kusimba (Eds.) *East African archaeology: Foragers, potters, smiths, and traders* (pp. 99–115). Philadelphia: University of Pennsylvania Museum of Archaeology and Anthropology.
Kusimba, C. M., Killick, D. J., & Cresswell, R. G. (1994). Indigenous and imported metals at Swahili sites on the coast of Kenya. In S. T. Childs (Ed.) *Society, culture and technology in Africa*. Philadelphia: MASCA, University of Pennsylvania Museum of Archaeology and Anthropology.
LaViolette, A. (2008). Swahili cosmopolitanism in Africa and the Indian Ocean world, AD 600–1500. *Archaeologies, 4* (1), 24–49.
LaViolette A., & Fleisher J. (2005) The archaeology of sub-Saharan urbanism: Cities and their countrysides. In A. B. Stahl (Ed.), *African Archaeology: A critical introduction* (pp. 327–352). Oxford: Blackwell.
Levtzion, N. (1973). *Ancient Ghana and Mali* (Vol. 7). London: Methuen.
MacEachern, S. (1993). Selling the iron for their shackles: Wandala-Montagnard interactions in Northern Cameroon. *Journal of African History, 34*, 247–247.
Mackenzie, J. M. (1975). A pre-colonial industry: The Njanja and the iron trade. *Nada, 11*(2),200–220.
Maggs, T. (1992). My father's hammer never ceased its song day and night': The Zulu ferrous metalworking industry. *Southern African Humanities, 4*, 65–87.
Mamadi, M. F. (1940). The copper miners of Musina. In N. J. van Wamelo (Ed.), *The copper miners of Musina and the early history of the Zoutpansberg*, (pp. 81–87). Pretoria (Department of Native Affairs, Union of South Africa): Ethnological Publications VIII.
Mauss, M. (1954). *The gift: Forms and functions of exchange in archaic societies* (No. 378). New York: Norton.
McNaughton, P. R. (1993). *The Mande Blacksmiths: Knowledge, power, and art in West Africa*. Bloomington: Indiana University Press.
Miller, D. (1996). *Tsodilo jewellery*. Cape Town: University of Cape Town Press.
Miller, D. (2000). 2000 years of indigenous mining and metallurgy in Southern Africa—a review. *South African Journal of Geology, 98*, 232–238.
Miller, D. (2001). Metal assemblages from Greefswald areas K2, Mapungubwe Hill and Mapungubwe Southern terrace. *The South African Archaeological Bulletin, 56*, 83–103.

Miller, D. (2002). Smelter and smith: Iron Age metal fabrication technology in Southern Africa. *Journal of Archaeological Science, 29*(10), 1083–1131.

Miller, D. E., & Hall, S. L. (2008). Rooiberg revisited–the analysis of tin and copper smelting debris. *Historical Metallurgy, 42*(1), 23–38.

Miller, D. E., & Van der Merwe, N. J. (1994). Early metal working in sub-Saharan Africa: A review of recent research. *Journal of African History, 35*(1), 1–36.

Miller, D., Killick, D., & Van der Merwe, N. J. (2001). Metal working in the Northern Lowveld, South Africa AD 1000–1890. *Journal of Field Archaeology, 28*(3–4), 3–4.

Mitchell, P. (2005). *African connections: Archaeological perspectives on Africa and the wider world*. Walnut Creek: AltaMira.

Mudenge, S. I. (1974). The role of foreign trade in the Rozvi empire: A reappraisal. *Journal of*

Oddy, A. (1984). Gold in the Southern African Iron Age. *Gold Bulletin, 17*(2), 70–78.

Nixon, S., Rehren, T., & Guerra, M. F. (2011). New light on the early Islamic West African gold trade: coin molds from Tadmekka, Mali. *Antiquity, 85* (330), 1353–1368.

Ogden, J. (2000). Metals. In P. Nicholson & I. Shaw (Eds.), *Ancient Egyptian materials and technology* (pp. 148-176). Cambridge: Cambridge University Press.

Phillipson, D. W., & Fagan, B. M. (1969). The date of the Ingombe Ilede burials. *The Journal of African History, 10*(2), 199–204.

Rehren, T., Belgya, T., Jambon, A., Káli, G., Kasztovszky, Z., Kis, Z., & Szőkefalvi-Nagy, Z. (2013). 5,000 years old Egyptian iron beads made from hammered meteoritic iron. *Journal of Archaeological Science, 40*(12), 4785–4792.

Reid, A., & MacLean, R. (1995). Symbolism and the social contexts of iron production in Karagwe. *World Archaeology, 27*(1), 144–161.

Robion-Brunner, C, Serneels, V, & Perret, S. (2013). Variability in iron smelting practices: Assessment of technical, cultural and economic criteria to explain the metallurgical diversity in the Dogon area (Mali). In J. Humpris & Th. Rehren (Ed.), *The world of iron* (pp. 257–265). London: Archetype.

Sasson, H. (1964). Iron-smelting in the hill village of Sukur, North-Eastern Nigeria. *Man, 64*, 174–178.

Scheel, B. (1989). *Egyptian Metalworking and Tools*. Oxford: Shire Publications.

Schmidt, P. R. (2009). Tropes, materiality, and ritual embodiment of African iron smelting furnaces as human figures. *Journal of Archaeological Method and Theory, 16*(3), 262–282.

Scott, D. A. (1991). *Metallography and microstructure of ancient and historic metals*. Oxford: Oxford University Press. (Getty Conservation Institute; J. Paul Getty Museum, in association with Archtype Books).

Shaw, T. (1970). *Igbo-Ukwu: An account of archaeological discoveries in Eastern Nigeria* (Vol. 2). Evanston: Northwestern University Press.

Shinnie, P. L. (1985). Iron working at Meroe. In R. Haaland & P. L. Shinnie (Eds.), *African iron working—Ancient and traditional* (pp. 28–35). Oslo: Norwegian University Press.

Stanley, G. H. (1931). Some products of native iron smelting. *South African Journal of Science, 28*, 131–134.

Stayt, H. A. (1931). *The Bavenda (No. 58). International Institute of African Languages & Cultures*. Oxford: Oxford University Press.

Tamari, T. (1991). The development of caste systems in West Africa. *Journal of African History, 32*, 223.

Thondhlana, T. (2013). Metalworkers and smelting precincts: Echnological reconstructions of second millennium copper production around Phalaborwa, Northern Lowveld of South Africa. Unpublished Doctoral dissertation, University College London, London.

Thondhlana, T. P., & Martinón-Torres, M. (2009). Small size, high value: Composition and manufacture of second millennium AD copper-based beads from northern Zimbabwe. *Journal of African Archaeology, 7*(1), 79–97.

Todd, J. A. 1985. Iron production by the Dimi of Ethiopia. In R. Haaland & P. L. Shinnie (Ed.), *African iron working-ancient and traditional* (pp. 88–101). Oslo: Norwegian University Press.

Vansina, J. (1969). The bells of kings. *The Journal of African History, 10*(02), 187–197.

Warnier, J. P., & Fowler, I. (1979). A nineteenth-century Ruhr in Central Africa. *Africa, 49* (04), 329–351.

Chapter 6
The Social Role of Metals

Introduction

The utilitarian, aesthetic and ceremonial values of metals are a pivot on which not just quotidian activities but also luxuries consumed by humanity were anchored since the advent of metallurgy whether in Africa, Latin America or Eurasia. Around c. 5500–6000 years ago, copper ores were used as ornaments in the Chalcolithic period of the Middle East and adjacent regions (Hauptmann 2007). Since then, metals have played an increasing role in the articulation of value systems, accumulation of wealth, and innovation and knowledge transfer. As such, metals occupy a privileged place within societies of the preindustrial period.

Diverse opinions prevail among researchers when it comes to the question of whether the adoption and subsequent entrenchment of metallurgy in preindustrial societies was a revolutionary change. For decades, the Childean view that the adoption of metallurgy caused a social revolution in Western Eurasia was influential (Childe 1930). This academic footing is now being challenged by positions which argue that the impact of metallurgy was gradually felt. Whether in Egypt, West or Southern Africa, metallurgy only increased in scale sometime after its introduction. It was only after a lengthy period—in the view of some, on a scale of millennia—that humanity's dependence on metals intensified. This anti-revolution school further criticizes Childe and his ardent students for embracing a presentist view which projects the all-encompassing might of metallurgy today deep into the past (Killick and Fenn 2012). Whatever the case may be, this contestation around the revolutionary and non-revolutionary consequences of metallurgy passes fascinating comments not just on the spatial and temporal continuities and changes across human history, but also on the scale and evolution of techno-social systems in the preindustrial world.

By consequence of valuation, access to metals, or lack thereof, became a tool for creating social differentiation and inequality in society. Metallurgy became entrenched in the evolution of craft specialization in predynastic Egypt with the rulers and those who controlled metal production, distribution and use becoming wealthier and concomitantly more powerful than those without (Childe 1930). After some time, the efficiency of metal tools in domains such as agriculture made food production comparatively easier, while metal weapons enabled various polities to defend and expand their territories. In the long term, urbanization promoted settlement aggregation in some but not all metalworking regions.

Because the distribution of ores is geologically specific and uneven, trade between resource-rich and resource-deficient areas understandably emerged. Therefore, local, regional and long-distance trade is one of the direct consequences of the value which metals had in society. As metal sources became depleted, societies sought access to new sources, initiating proto-forms of globalization anchored on not just land links but also sea links. For example, from the Early Bronze Age, Egypt was part of the trade and exchange relationship in the Middle East. By 2000 BC, Egypt had expanded into Nubia in search of gold. This continued until Roman times when Meroe used to supply iron to the Romans. Sub-Saharan Africa also supplied iron and gold metal to the nascent global system of the Islamic world via the trans-Saharan trade and the East African coast-based trade particularly after the first-millennium AD (Nixon 2009; Nixon et al. 2011). With the advent of trade via the Atlantic coast from the 14th-century AD onwards, African metals paid for valuables initially in Europe and later for spices in India. Trade in metals became invariably associated with the development of, and transfer of, diverse value and knowledge systems. Fighting over control of metals sources spawned conflicts and cleavages that often resulted in mobility, migration, war and even colonization.

General Impact of Metals in Society

Throughout the world, metallurgy for a long time coexisted and only gradually replaced stone as the major medium for manufacturing tools for a variety of purposes. This conservatism is understandable given that for a greater part of humanity, stone was the primary raw material for making tools. In terms of materiality, metals hold several advantages over stone because of their ductility, ability to undergo plastic deformation, malleability and strength (Scott 1991). When realized, these qualities dictated that metals were used in a variety of domains, as underlined by the values with which humanity associated them. In the pre-Columbian New World, stone remained the basic raw material for tools and weapons while metals were enjoyed by elites. The range of the values of metals varied from the more mundane ones to the higher order ones as dictated by rarity and scarcity.

It has been argued that when agricultural communities adopted metallurgy, the technology had a beneficial effect through the provisioning of efficient tools. Although there are instances of cultivation with polished stone axes in West Africa (for example the Kintampo complex in Ghana and the Neolithic in Gabon; Anquandah 1993), it seems highly improbable that swidden agriculture in the savanna woodlands (especially the miombo of Central and East Africa) would have been possible without iron (Killick 2014). Metal tools such as hoes and axes made it far easier to cultivate the land and to clear the vegetation which in turn opened up more land for cultivation (Phillipson 2005, p. 214). The burning of the trees produced ash, which increased the fertility of the land. These activities resulted in increased yields, which made it easier to sustain growing populations. In Southern Africa, the advent of metallurgy supported agriculture, opened up the land which drove away tsetse flies making it possible to practice cattle husbandry (Mitchell 2002). However, it must be noted that by no means did all communities adopt metals for food production. Some Aksumites used stone implements for cultivation until the early first-millennium AD because possession of metal correlated with class (Phillipson 2005). Another parallel can be found in the New World where before Columbus, various agricultural communities used stone for agriculture and metals for luxuries (Killick and Fenn 2012). Another area in which metallurgy supported the local economy and productive base in sub-Saharan Africa is within the sphere of hunting. Metallurgy facilitated the production of spears, arrowheads and axes that were more effective in hunting elephants for ivory, a very significant commodity in trade and exchange relationships between Africa and Eurasia, particularly after the mid-first-millennium AD when various parts of Africa were integrated into long-distance trade via the Sahara and the Indian Ocean. However, the advantages of metal over stone were only realized over time and were thus not immediate.

Besides these utilitarian tasks, metals were highly valuable in making ceremonial and decorative items. Various bronze and gold castings adorned ancient Egyptian temples and palaces. The famous Igbo Ukwu bronzes dating between AD 900 and 1200 (Shaw 1970) and the Benin bronzes made from the 16th-century AD onwards were used to express power and to commemorate deceased kings and were an inimitable expression of the Oba's power and authority (Herbert 1984). In Great Zimbabwe (AD 1100 to 1550; Fig. 6.1), archaeologists recovered impressive bronze spearheads and iron gongs (Fig. 6.2) which, as elsewhere in Central and West Africa were associated with the power of kings (Vansina 1969). Copper, bronze, gold and iron were used to make beads and bangles for personal adornment (Miller 1996). The values associated with colours and rarity of these metals also created a distinguishing factor between elites and non-elites (Smith 1981). For instance, in Ancient Egypt and Nubia, gold was an elite metal hardly enjoyed by the commoners. Metal was subjected to rigorous bureaucratic control. In contrast, commoners in Southern Africa produced gold that was traded by the Rozvi state (AD 1680–1850) and even exchanged it for glass beads which they gave to rulers are tribute. (Mudenge 1974). As such, they were free from the bureaucratic control, which gave them some degrees of freedom when compared to ancient Egypt and Nubia.

The power of the hugely successful Asante kingdom (after AD 1600) in modern-day Ghana was symbolically vested in the Golden Stool which became an emblem of the polity (Garrard 1980, pp. 63–64). The Asante royalty also consumed significant amounts of gold, which seemed as a high-status metal (Garrard 2011). In Southern Africa, gold was a universal feature of centres of power such as Great Zimbabwe, Khami and Mapungubwe (Figs. 6.3 and 6.4) but its production was

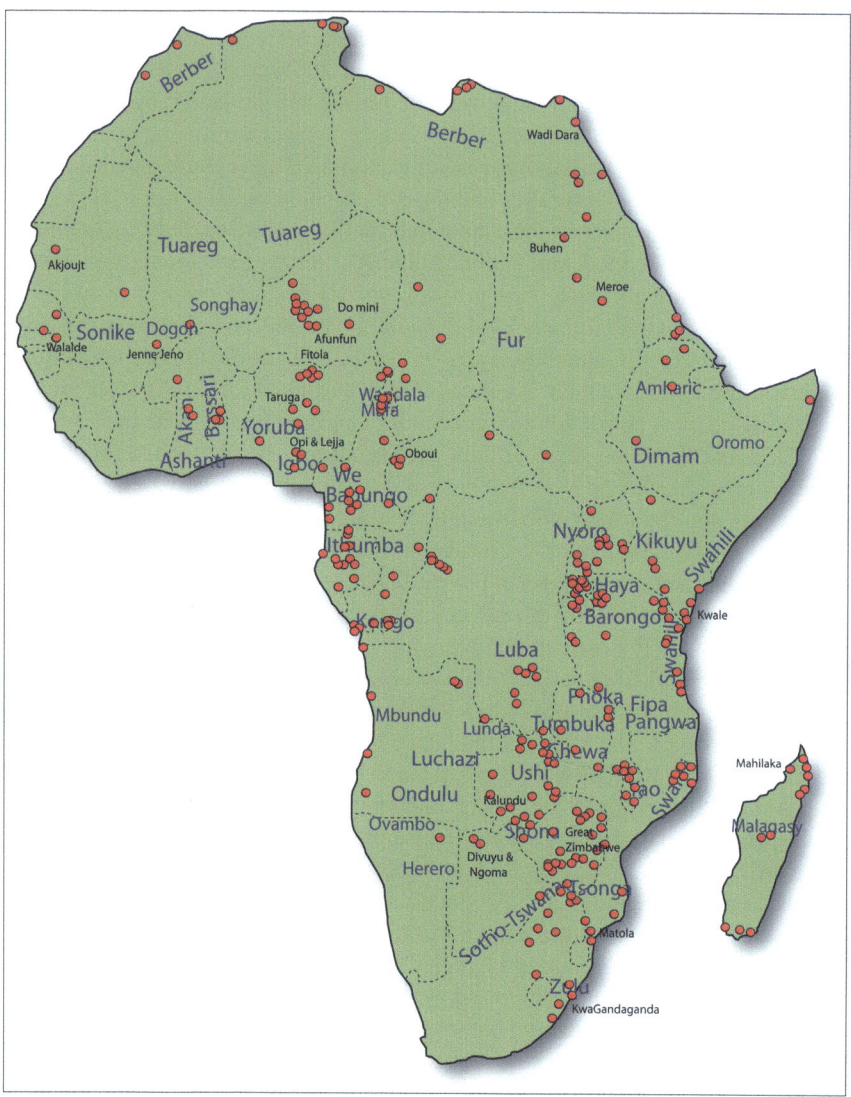

Fig. 6.1 Location of metal working groups and places discussed in the text

not under very strict bureaucratic control. Besides making symbolic objects and ornaments for personal adornment, thin sheets of gold were attached to wooden posts inside royal abodes at places such as Great Zimbabwe. The gold sheeting was designed to enhance the aesthetic appeal of places of power. In contrast, there is almost no record of the presence of gold at commoner settlements in much of Africa. It appears that copper was the preferred metal (Herbert 1984), but it must be borne in mind that little work has been directed at commoner residences.

Fig. 6.2 Iron gong recovered from Great Zimbabwe. Natural History Museum, Bulawayo (Photo: Author)

Fig. 6.3 Gold leaf bowl from Mapungubwe, Southern Africa. It is believed that the leaf was attached to a wooden core which has since decayed (Miller 2001). (Source: University of Pretoria Mapungubwe Collection)

Fig. 6.4 Mapungubwe golden rhino made of gold leaf (size, c. 10 cm). It symbolized the "majesty" of kingship. (Source: University of Pretoria Mapungubwe Collection)

From predynastic Egypt to Ptolemaic Egypt and Nubia, gold foil decorated the interior of temples, tombs and royal palaces and was thus a high-status metal from its discovery onwards. However, there are some African communities particularly those in Central Africa where copper and not gold was an important and high-status metal (Herbert 1984). Indeed, Sanga burials dating from the first-millennium AD up to the mid-second-millennium AD in the Upemba Depression had varying quantities of copper (Bisson 2000). Indeed, Cline (1937) argues that some communities in modern-day Angola told the Portuguese that alluvial gold was abundant in some of the major rivers but that they did not value gold when compared to copper.

The role of metals in the social nexus between societies is illustrated by MacEachern's study of power and political relations between the Muslim Wandala state and populations on the margins, particularly those known as *montagnards* or non-Muslims. MacEachern (1993) combined historical observations with archaeological work to document these relations. Non-Muslims occupied the Mandara massif while the Wandala lived on the plains. MacEarchern (1993) noted significant demographic shifts marked by increased *montagnard* occupation of the mountain from the early second-millennium AD onwards. These mountain groups were organized in small patrilineages with densities of between 50 and 200 people per square km. They specialized in iron production aimed beyond local requirements. The *montagnards* traded iron ore and iron blooms in exchange for foodstuffs and other necessities brought by the Wandala. The Wandala, in turn, traded the iron with Bornu and other leading centres of the time. Ironically, the guns and cavalry obtained from this trade enabled Wandala people to capture *montagnards* as slaves, resulting in the latter being shackled by iron they had produced (MacEachern 1993). This example shows the dynamic and changing relationships in society that cascaded from iron production, its exchange and associated consequences. Indeed, it has been suggested that in other regions of West Africa such as Bassar, the trans-Atlantic slave trade resulted in increased iron production, although over time, associated insecurity is supposed to have decreased production levels. However, as the *montagnard* and Wandala case study shows, the situation was very complex because overall the trade could be beneficial and at other times toxic. It is the beneficial aspect that fuelled iron production—basic foodstuffs and other luxuries and necessities were important—but those who accepted iron turned the product against its producers as they came back to raid for slaves.

Metals, Sociopolitical Complexity and Urbanism

Metallurgy is believed to be one of the key drivers of sociopolitical complexity and concomitant urbanization (Chirikure 2007). These related ways of societal organization immensely transformed the African sociopolitical and physical landscape. Sociopolitically complex societies are often ranked, have evidence of craft specialization, division of labour, advanced subsistence and economic systems, monumental architecture, religion, large populations and writing (Carneiro 1967;

Renfrew and Cherry 1986). Urbanism is a form of social organization in which specialized centres provide and receive specific services from hinterland areas (LaViolette and Fleisher 2005). The appropriateness of some of these indicators in non-western communities has been questioned by scholars such as Connah (2001) and McIntosh (1999) who argue that a contextual approach is required to best understand African varieties of urbanism.

Although the physical evidence for sociopolitical complexity and urbanism differ across sub-Saharan Africa (see Mitchell and Lane 2013), the most widely cited cases of urbanism are those of the West African Sahel, the West African rain forest and its fringes, the middle Nile in the Sudan, the Ethiopian and Eritrean mountains, the Swahili settlements of East Africa, the Zimbabwe plateau, Tswana towns of 18th- and 19th century Southern Africa, and the capitals of states in the Upemba depression in Central Africa (Connah 2001). Historical evidence suggests that there was a tendency for population to agglomerate in much of sub-Saharan Africa, such that by the close of the 19th century, larger areas of precolonial Africa were quintessentially urban in character (Fletcher 1993; Fig. 6.5).

The quintessential sub-Saharan urban centre has distinguishing characteristics which vary from place to place, but the agglomeration of varying numbers of people in one area seems to be the common denominator. Often, urban centres in Central and West Africa were distinguished by settlement clusters demarcated by streets. In Sudanic West Africa, perimeter walls of mud brick often encircled the urban settlement (Ogundiran 2005). In the forest zone, major earthworks created a network of cells in which a dispersed urban population lived. In Southern Africa, the Zimbabwe culture urban centres such as Mapela, Mapungubwe, Great Zimbabwe and Khami were heavily built up and consisted of impressive dry-stone walled enclosures and platforms constructed using the precise placement method without any binding mortar (Chirikure et al. 2013). Metallurgical specialization and division of labour is an all pervading characteristic of most urban centres in Egypt, Nubia and Aksum, but elsewhere in West Africa there is little evidence of specialist metal production at urban centres (for exception see Garrard's 2011 discussion of specialist gold artisans employed by the Asante kings). In Southern Africa, Tswana towns such as Marothodi were specialist metal-producing towns that worked copper, iron and tin bronze exchanging the metals and alloys locally and regionally (Hall et al. 2006). Although debatable, it has been argued that Great Zimbabwe was serviced by a large-scale metallurgical industry at places such as the nearby Chigaramboni Hill (Ndoro 1994).

Overall, the bureaucratic control of metal production and use in Egypt and Nubia when compared to a contrasting lack of rigorous control in sub-Saharan Africa questions the depth of arguments that link monopoly over trade in gold with the rise of urban centers and more complex state systems in the former. It seems that the power base was rooted in something else other than control of gold trade. Control of the land, people, cattle and general fertility are among the possibilities (Mudenge 1974).

In order to evaluate metallurgy's role in the emergence of sociopolitical complexity as broadly defined, it is important to briefly sketch the pathways to urbanism in Africa. It seems that different regional histories motivated for differences in

Fig. 6.5 Location of prominent urban centers in Africa

pathways to complexity across much of Africa (McIntosh 1999). In West Africa, archaeologists have advanced the theory that the advent of food production and not metallurgy created itinerant pastoral elites who built the megaliths of Dar Tichitt in Southern Mauretania (MacDonald 1998). Similar processes where stone-using food-producing elites built large monumental structures were also reported at Zilum on the Chad plain (Magnavita et al. 2006). While these cases are a manifestation of heterarchy or some form of social stratification, it is not clear whether the societies

Fig. 6.6 Indian Ocean cowrie and sea shells made their appearance in Southern Africa from AD 700 onwards (Photograph: Author)

that built them can be described as fully urban (McIntosh 1999). However, the presence of gigantic megaliths implicates an ability to mobilize and control pools of labour by one section of the population. Holl (2009) argues that, even if this early sociopolitical complexity was not associated with metallurgy, the prevailing evidence suggests that in much of Africa metallurgy became the oil that lubricated state formation and urbanization.

Southern Africa paints a vivid contrast on the canvas of the role of metallurgy in the beginning and flourishing of sociopolitical complexity. In this region, agriculture appeared simultaneously with iron and copper metallurgy early in the first-millennium AD. Perhaps the region benefited from its "late adopter" advantage, skipping experimenting with metallurgy in the face of a deeply entrenched stone making technology. Yet even in Southern Africa, it took between two and three centuries for sociopolitical complexity to emerge following settlement of the region by iron-using agriculturalists (Chirikure et al. 2013). Pwiti (1996) identified four stages in the organization of Southern African communities between the mid-first-millennium AD and the early second-millennium AD. In the first stage (ca. AD 300 to 500), heterarchically organized early farming communities occupied dispersed villages, but metalworking was a prominent feature of such societies. The second stage, extending from the 7th- to the 8th-century AD, saw the introduction of external trade. Archaeologically, there was a shift in production towards goods with long-distance exchange value (Pwiti 1996). The third stage, from about the 9th-century AD, sees an increase in the volume of external trade and is characterized by villages which begin to show evidence of ranking and social differentiation. Shells from the coast also start to make their appearance (Fig. 6.6). The fourth and last stage sees the establishment of state structures such as Mapungubwe, Mapela and Great Zimbabwe. Fagan (1969) argued for the presence of wide regional trade networks involving ores and metals in Central and Southern Africa since AD 700. The size of some of the early to mid-first-millennium villages such as Swart Village (Chirikure 2007) suggests some form of ranking, but African archaeology has yet to develop indicators of ranking where locally produced goods such as cattle are involved. Quite interestingly, there is no evidence of large-scale metal production in Southern Africa at any period that has implications for organization of production. If the production was carried out at villages scattered across the landscape with the surplus being siphoned through various mechanisms (Mudenge 1974), it is unlikely that large-scale remains will ever be found. However, this does not negate the fact that metals formed a critical element of local and long-distance trade. Again, this challenges the beliefs that even centralized states such as the Rozvi (AD1680–1850), Great Zimbabwe (AD 1100 to 1550) and Mapungubwe (AD1220–1290) monopolized trade in metals.

In sub-Saharan Africa, the story of iron smiths and metalworkers who forged new states and brought civilization is widespread across the breadth and length of the region from the first-millennium AD until the 19th-century AD (Bocoum 2006; Chirikure 2007; de Maret 1985; Holl 2009; Humphris and Iles 2013; Killick 2009; Kim and Kusimba 2008). Holl (2009) presents interesting observations relating to the contribution of metallurgy to state formation in West Africa. There are some indications that radical change in the social status of metalworkers took shape in the first half of the second-millennium AD in West Africa. For example, the founding dynasty of Takrur, the Jaa-Ogo, was of iron-producer extraction. An exclusive control of the craft and its concomitant esoteric knowledge, was the core reason for the Jaa-Ogo's accession to kingship. Takrur was invaded and conquered by a Soninke army from the neighbouring kingdom of Ghana in the 11th-century AD, prompting the Jaa-Ogo dynasty to lose political power in the process (Holl 2009). These observations have resonance with those made by Huysecom and Augustoni (1997) who argued that the Dogon kings of Mali were reputed ironworkers. Tamari (1991) studied the development of endogamous occupational groups or castes in West Africa. Ironworkers and potters belonged to the same caste who inter-married and were often on the low levels of the social ladder. These endogamous castes were totally absent in Southern Africa. Tamari (1991) argues that ironworkers were politically powerful in West Africa until the formation of the state of Mali, which engineered their marginalization. This seems to reach some convergence with Holl's (2009) argument that in Senegal area, the marginalization of ironworkers was a by-product of Islamization which took place from the 9th-century AD onwards (see also McNaughton 1993).

The positive correlation between political power and metallurgy was also a characteristic of first-millennium and second-millennium AD Central and East African communities. Urban centres such as Buganda were supported by extensive iron production industries which produced weapons and utilitarian tools while the Bachwezi used iron objects during royal investiture (Reid and MacLean 1995). Perhaps the most unequivocal case is that of the Luba and Lunda kingdoms whose origin myths make it explicit that their founders were great metal smiths (de Maret

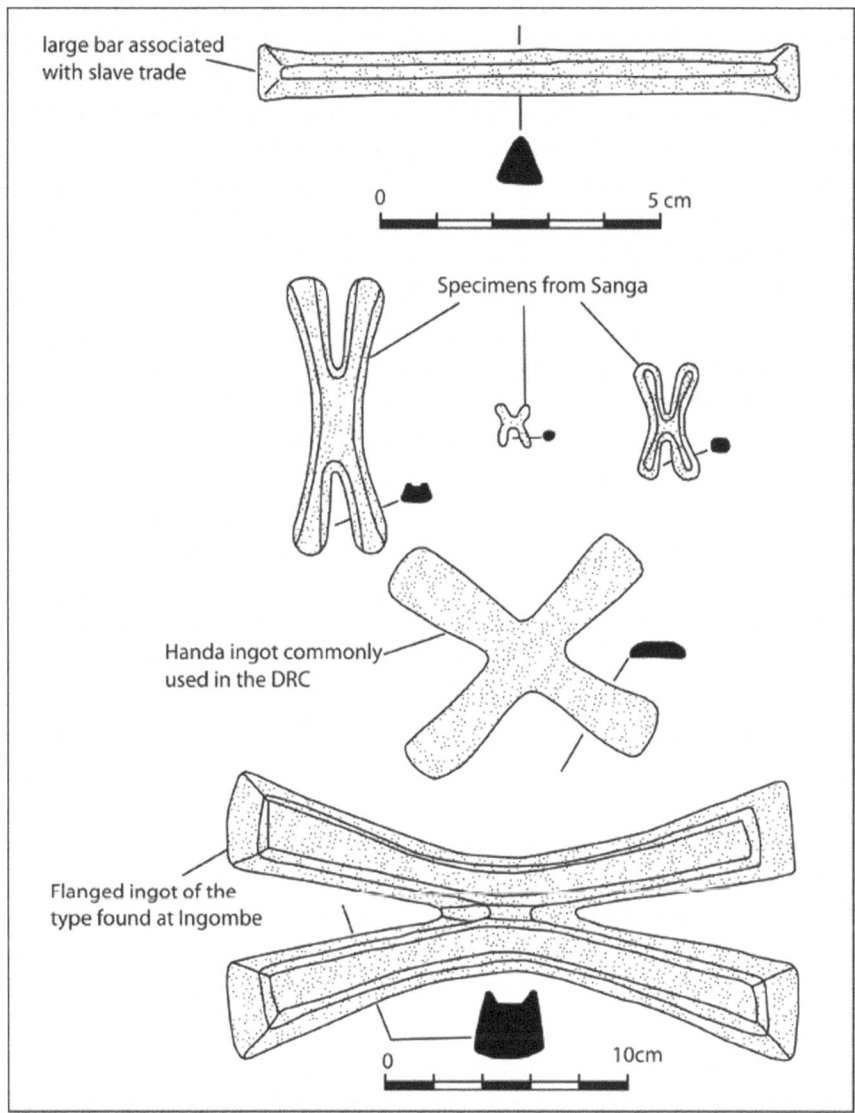

Fig. 6.7 Ingots from burials excavated from the Upemba depression in the modern-day DRC. (Redrawn from Bisson 1982, p. 131; Fig. 6.2)

1985). In these kingdoms, royal inauguration involved symbolic appropriation of the occult powers of the ironworker, with kings being symbolically transformed into positions of authority. According to de Maret (1985), this practice is for most of the time archaeologically detectable in Central Africa. For instance, excavations of burials in the Upemba depression uncovered people buried with accompanying ingots (Fig. 6.7), smiths' tools/symbols demonstrating that metallurgy was a prestige technology used as leverage to political power.

In Southern Africa, the 17th-century Njanja (part of group labelled Shona in Fig. 6.1) presents one of the best and most remarkable examples in which renowned metal smiths used their expertise in metal working to establish a large chiefdom. The Njanja people migrated from Sena in Mozambique and settled in modern-day Central Zimbabwe sometime in the 17th century (Chirikure 2006; Mackenzie 1975). Oral traditions claim that Neshangwe, the founder of this chiefdom, was a reputed ironworker whose skills attracted followers precipitating the establishment of a chiefdom leveraged by iron production. The Njanja reorganized their production by employing a shift system of labour and by being open to use of female labour when some groups seemed to shun it. The result was a distribution network covering a 200-km radius (MacKenzie 1975). This and other examples summarized above demonstrate that metallurgy was an important tool for social engineering. States were made and unmade in no small part by the symbolic and practical importance of metallurgy. In fact, metal participated in relations of coercion.

Trade in metals, particularly gold, has been touted as one of the most significant ingredients in emergence of hierarchical societies and early state systems. In Southern Africa, it is believed that gold production was the pulse for the florescence of state systems based at Khami, Mapungubwe and Great Zimbabwe (Killick 2009). Gold featured heavily in trade and exchange relationships between the Zimbabwe Culture states and the Indian Ocean trading system. Ruling elites controlled this trade and amassed substantial amounts of wealth, which was important in giving them political power. Not surprisingly, gold was an elite metal, only associated with centres of power such as Bosutswe, Mapungubwe and Great Zimbabwe to mention but a few. The available evidence suggests that gold was produced in the hinterland areas, but commoners were free to trade it for various commodities. In some cases however, although production was not controlled, once in the hands of elites its external distribution and internal consumption was (Phimister 1974; Mudenge 1988). Proceeds from trade in gold were then invested in building monumental architecture and other elite displays of power. The trade in gold, iron and other resources such as ivory brought in bronze, glass beads and other commodities which were exploited by elites to cement their power. According to Killick (2009), the introduction of bronze and gold represent the adoption by the new Southern African elites of an alien value system, in which the golden colour of gold and bronze supplanted the red of copper for personal ornamentation among emerging elite.

Although gold trade was important in state formation, the gold trade stimulus hypothesis is fraught with inconsistencies that render it highly problematic (Mudenge 1974). The first is that although gold trade was important, the dominant thinking makes it appear as if the impact of gold trade was instantaneous when in actual fact it was gradual. This is supported by the observation that even though trade between Southern Africa and the Indian Ocean started around AD 700, gold only becomes visible archaeologically at the beginning of the second-millennium AD when elements of ranking and complexity had already developed (Pwiti 2005). Trade in gold only intensified pre-existing processes involving control over iron production, fertility, land, ritual and other local ideologies (Chirikure 2007). In West Africa, the rise and florescence of ancient Ghana, Mali and Songhai is intricately associated with

Fig. 6.8 Iron projectiles used by the Buluba of the Katanga region of the Democratic Republic of Congo. Natural History Museum of Zimbabwe, Bulawayo

trans-Saharan gold trade channeled to the Islamic world via entrepots such as Tewdagoust (Levtzion 1973; Nixon et al. 2011). Although gold trade was important, it is now generally accepted that local factors and the contributions of other metals such as iron have been lamentably downplayed (Mudenge 1974). Therefore, the argument that certain ranked polities seized control of the earliest gold trade and very rapidly became states does not appear to be convincing in contexts where there is no strict bureaucratic control over its production as in Nubia and Egypt. The problem becomes one of enforcing the control and monopoly given the capacity challenges facing many developing states. As Mudenge (1974) argues, the kings levied tribute in glass beads, cloth, gold, iron and even cattle and for commoners to obtain exotic goods they had to trade in gold and other resources, thereby weakening hypothesis that elites at Mapungubwe, Mapela, and Great Zimbabwe monopolized long-distance trade.

The agglomerations of large populations at urban centres and the associated hinterland areas created problems of control for the rulers and elites. As such, one of the major contributions of metallurgy to African social engineering was in the provision of weapons for defending territorial integrity as well as for territorial expansion (Fig. 6.8). For example, armed with mass produced (Maggs 1992) short stabbing spears, Shaka's 19th-century army was instrumental in building a substantial urban centre as well as a state in the KwaZulu-Natal region of South Africa (see Fig. 6.1). The army was used to gain the allegiance of different groups of people. In West Africa, the well-respected army of Samori Toure was also armed with iron weapons, although Samori later had gun smiths who imitated European imports thereby demonstrating ingenuity and flexibility (Adu-Bohen 2006). Going back in time, Portuguese records on the Zimbabwe plateau mention that Mutapa armies were armed with iron spears which were used to maintain territorial integrity. The Portuguese were even expelled from the plateau in the 17th century by a force of Changamire armed with those spears (Mudenge 1988). In Central and West Africa, the armies of the Mbanza kingdom of Kongo were also armed with spears just as were the Luba and Lunda.

Therefore, as a socially embedded technology, metallurgy had an all pervading influence in society. Control of the knowledge of metallurgy was important to economic and political centralization, just as the tools and weapons were important in subsistence, economic and defence spheres. The representation and materialization of power in objects provided a link between royal ancestors, political authority and the aesthetic and physical properties of metals. As a socially embedded technology, the adoption and entrenchment of metallurgy had numerous socioeconomic and political permutations, particularly in the emergence of state systems.

Metallurgy, Culture Contact (Interaction), Proto-Globalization and Technology Transfer

The values attached to metals in antiquity, when coupled with their differential availability, were some of the principal motivators for initiating local, regional, as well as long-distance cultural contact on the land and sea. The resulting trade and exchange network connected people of different cultures and value systems, leading to complex intersections of local and non-local systems from early on. For instance, the Romans exported their values while adopting those of others. During the golden age of Islamic civilization, most of the Old World adopted some technologies and values from the Middle East. This process was responsible for the Islamization of West and East Africa from early on. From the 19th century onwards, Europeans successfully exported industrial capitalism to all parts of the globe, resulting in intensified cultural contact, technology transfer and biological and cultural exchanges.

This contact took place locally, regionally and internationally. Although not always emphasized in African archaeological literature, presumably because of the difficulty in identifying items traded locally, there existed a very strong and vibrant local and regional trade involving metals throughout sub-Saharan Africa. Processed metals and alloys as well as raw ore were widely traded in the subcontinent and associated with both biological and technological exchanges. Historical and ethnographic sources allude to the existence of an intricate local trading system involving copper and iron at Musina and Phalaborwa in Northern South Africa after AD 1000. The famous Venda *musuku* and Phalaborwa *lerale* ingots served as one of the primary forms in which copper was circulated in the lowveld and beyond. Venda kings transformed the once independent Lemba copper workers into attached specialists who produced metal for the king (Stayt 1931). More importantly, there also existed a thriving trade in diamond-shaped hoes produced in Phalaborwa but with a wide circulation covering adjacent parts of Zimbabwe and Mozambique after the second half of the second-millennium AD. The technological style of making diamond hoes (Fig. 6.9) as well as the *musuku* and *lerale* ingots became a "brand" associated with metal workers of Northern South Africa in the second-millennium AD. Surprisingly, there was little imitation, for the Shona across the Limpopo preferred oval-shaped hoes, demonstrating that social differentiation can, to some extent, be read from the typology of metal objects. The metal was exchanged for cattle, grain

Fig. 6.9 Diamond-shaped hoes produced by Phalaborwa smiths in Northern South Africa between c. AD 1600 to 1900. The hoes were used as currency (Photo credit author)

and other local commodities which leave little in terms of material remains. This invisibility explains why the long-distance contact that brought durable and visible commodities such as beads is emphasized in importance over local trade. In Southern Cameroon, a vibrant trade involving very thin iron bars known as bikies was also practised. These strips were a form of currency and a medium of exchange with different bundles possessing different values ranging from a wife, a cow to basic foodstuffs (Bohannan 1955; Ringquist 2008). As with Southern Africa, it is difficult to archaeologically study this trade owing to the poor survival rate of some of the commodities involved.

In terms of regional trade, one of the most important, but poorly understood cases relates to the exchange of copper and iron between Southern Zambezia and Central Africa in the first- and second-millennium AD. The very distinctive copper ingots called Katanga crosses which were made in this part of Africa were found at Great Zimbabwe and other places to the north (Garlake 1973). However, little compositional or isotopic work has been conducted on the Katanga crosses in Southern Africa to evaluate if they were imitations. There also existed iron gongs which were recovered from Great Zimbabwe (AD 1100 to 1550) and certainly originated through contact with Central Africa. Possibly, cattle, grain and other local commodities formed part of this exchange network.

Another important regional trade involved metallic tin in Southern Africa. Rooiberg in the Southern Waterberg (Fig. 6.10) is the only uncontested source of tin in preindustrial Southern Africa. Estimates by early mining geologists indicated that 18,000 t of ore had been mined preindustrially but interestingly there is very little in terms of settlement around Rooiberg (Chirikure et al. 2010). In contrast to this, various forms of tin ingots were recovered from numerous places north and northeast of Rooiberg such as Polokwane, Venda and among others Great Zimbabwe (Grant 1999) (Fig. 6.10). Geochemical and isotopic fingerprinting indicated that virtually all these tin ingots scattered throughout Southern Africa were made using Rooiberg tin. The mechanisms of this trade are unknown, but it is possible that it either was

Fig. 6.10 Location of Rooiberg in relation to capitals

down the line trade or a series of middleman traversed the landscape from corner to corner exchanging the metal for various merchandize. Middlemen such as Tsonga and Shona were famous in this regard (Mudenge 1988).

In West Africa, the increasing scale of production, particularly from the late first-millennium AD onwards, has strong implications for regional trade. De Barros (1986) argues that the Bassar iron trade was geared towards meeting the demands of an external market. At Sukur in Northeastern Nigeria, there was a semi-industrial process aimed at producing iron objects which were exported to Bornu and other centres (David and Sterner 1997; Sasson 1964). Equally, the montagnards occupying the Mandara Mountains of Northern Cameroon produced iron blooms which were used by Wandala to produce shackles (MacEachern 1993), thereby feeding into the demands of the slave trade. Another important large-scale iron production centre in Cameroon is that of the Ndop plain discussed by Warnier and Fowler (1979). The scale of production in all these areas was staggering and exceeded that produced at places such as Meroe (de Barros 1986). According to David and Sterner (1997), iron was the source of power in Sukur and adjacent areas.

This regional contact was also responsible for mobility, technology transfer and cross-borrowing. For example, contact between Central, Eastern and Southern Africa was responsible for the dispersal of techniques such as wire drawing (Bisson 2000). However, there was an element of conservatism too, for various regions continued to be associated with distinctive furnace and metal ingot types. It appears that each group was happy to receive metal and work it using locally acceptable methods. The low rate of technology transfer was most likely an outcome of specialization. Specialists produced their metal within culturally acceptable norms which often constrained the readiness with which they could welcome techniques from elsewhere. Without exporting ethnographic understanding to the deep past, it is debatable whether different technological styles diachronically and synchronically became bound up in processes of identification in relation to what we understand today as ethnicity or not. Often, metalworkers migrated to areas rich in ore supply. In the new areas, they inter-married locally resulting in genetic flows. A good example is provided by the Njanja who migrated from Northern Mozambique and established home in Central Zimbabwe (Mackenzie 1975). There is no reason why the proceeds from local and regional metal trade would not have been important in fostering societal transformations.

Besides localized and regional trade in metals, there existed a long-distance trade network that connected together interior communities with those resident at the coast and beyond. One of the oft-cited cases of long-distance interconnections relates to the Indian Ocean-based trading system that connected Southern and Eastern Africa and the Indian Ocean rim regions such as Persia, the Indian subcontinent, and regions afar. The Periplus of the Erythrean Sea mentions that metal objects were exported to East Africa by Greco-Romans c. 50AD. This trade became more visible from the 8th-century AD and involved iron, gold, ivory and other local commodities. From its advent until Vasco da Gama's voyage to India, this trade involved Swahili middlemen who were happy to travel into the interior to barter their commodities. At the same time, Shona middlemen known as *vashambadzi* obtained commodities from the Swahili and exchanged them in the interior for a profit (Mudenge 1988). In other contexts, the Shona travelled to the coast to obtain the goods which they exchanged locally. Al Masudi (cited in Summers 1969) reported that iron exported from East Africa was highly valued in the Indian Ocean world, together with gold and possibly copper (Kusimba 1999).

The success of this trade saw Greco-Roman merchants and later those from India, Arabia and later one Chinese expedition anchoring at East African ports with merchandize for trade from very early on. This brought in Indian glass beads, Persian glass beads, Chinese porcelain as well as alloys such as bronze, brass and leaded gun metal (Robertshaw et al. 2010; Stanley 1931). These metals were worked using local technologies of manipulating metals. It has been argued that long-distance trade introduced alien value systems which were appropriated and monopolized by the elites. For example, gold became a prestige metal together with bronze and the very rare imports of Chinese celadon and porcelain (Fig. 6.11), Persian wares and glass beads. Of all the imports, glass beads seem to dominate as they were recovered from both commoner and elite sites in large numbers. In fact the abundance of glass beads at elite and non-elite sites poses the question whether glass beads were status indicators or they were just an item of dress and decoration. Early on around, AD 700 when they were introduced, beads might have been prestige goods but by about AD900, they increased in frequency to the point of not being rare suggesting that their value may also have decreased. This is pertinent because Wood (2012) demonstrated that quantities of beads recovered from elite and non-elite areas in the

Fig. 6.11 Ming Dynasty porcelain from Zimbabwe housed at Iziko Museum, Cape Town. (Source: author)

Shashi-Limpopo do not differ much. This suggests that the influx of goods was not subject to royal control. In the historical period, many young men who were near marriage age could travel to the coast in search of beads or *chuma* to give to their prospective bride. Furthermore, *vashambadzi* or middlemen obtained large amounts of beads which were exchanged internally (Mudenge 1988). During the Portuguese period in Southern Africa (AD 1500–1900), markets or *feiras* were established in the interior at places such as Massapa (Baranda), Dambarare and Luanze. Direct trade took place at these centers while middlemen took the commodities to distant lands (Pikirayi 2001). It is possible that this may not apply to earlier times when the Zimbabwe state was likely more unified, but the absence of a formal bureaucracy makes it unlikely.

The other important question regarding value transfer is why Chinese porcelain was not popular in Southern Africa when it was present in substantive quantities at the East African coast (Horton and Middleton 2000). Perhaps this has something to do with the internalization of alien values. Although porcelains were highly valued, they may not have been widely accepted locally when compared to glass beads. It is unlikely that porcelain was so expensive to the extent of being prohibitive. Porcelain has mostly been found at elite sites but this is also where most research has been carried out (cf. LaViolette and Fleisher 2005). Garlake (1968) discusses the presence of porcelain from the Marcadoni claims in the very auriferous rich West Nicholson area of Southwestern Zimbabwe. Even at the elite sites, the porcelain and celadon fragments are few, and if Garlake's (1968) estimate is to be believed, not more than 90 fragments of the pre-Portuguese porcelains were recovered on the Zimbabwe plateau. When contrasted with the abundance of glass beads, cowrie and alloys such as brass and gun metal, it is clear that porcelains may have failed

Fig. 6.12 Regional and trans-continental connections between Southern and Western Africa and the trans-Saharan and Indian Ocean worlds (Base map: Google Images)

to appeal to local values in a way these other objects did. This is hardly surprising for Hall (1998) who suggested that even within a colonial frontier, imports struggled to replace domestic pottery in many activities. It may be possible that this was the same situation with porcelains. Although seemingly radical, this thinking is adequately supported by two independent but related observations. The first is that the Portuguese complained that hinterland communities shunned European glass beads in favour of the deeply entrenched Indo-Pacific beads (Wood 2012). Secondly, although the Portuguese brought chocolate on white porcelain as an item of trade, comparatively it does not seem to have gained wide local acceptance when compared to glass beads and guns for example. As a result, it is only concentrated at trading sites such as Dambarare (Pikirayi 2001). These points reinforce the observation that people consumed goods according to their own logics and prior preferences (Prestholdt 2008).

While these interconnections were taking place on the East African littoral, parallel developments were unfolding in West Africa where long-distance commercial traffic lubricated by trade in gold and slaves was deeply rooted (Fig. 6.12). As with Southern Africa, long-distance trade in West Africa exploited pre-existing localized and regional trade in regions such as the inland Niger Delta of Mali. This exchange brought alloys such as bronze and brass as well as glass beads. The 9–10th century AD site of Igbo Ukwu in Nigeria participated in this trade as evidenced by the recovery of burials with thousands of imported glass beads. As discussed above, Igbo Ukwu also yielded spectacular bronze artifacts produced through the lost wax method that were bound up in the materialization of power and ideology (Shaw 1970). Gold played a key role in the flourishing caravan-based trans-Saharan trade that enchained the Islamic world and West Africa. The successive states of ancient Ghana, Mali and Songhai extensively traded the gold from Bambuk, with merchants based at places such as Tewdagoust. This trade brought in salt and exotic goods such as glass beads as well as brass and other imported metals and alloys. The story of the legendary Mansa Musa (c. 1280–c. 1337), the Mali monarch who went on pilgrimage to Mecca carrying with him significant amounts of gold, is one of the best-known examples involving gold trade. So strategic was the control of the gold trade that hegemony over source areas was a question of life and death for these entities. However, there are successful states such as Kanem-Bornu near Lake Chad which, like many communities in Northern Nigeria and Northern Cameroon, did not participate in the gold trade (Garrard 2011). From the 14th century, West Africa became increasingly connected directly to Europe via the Atlantic littoral. This trade resulted in an influx of huge amounts of European alloys such as bronze and brass and metals such as iron. Gold, particularly that from the Gold Coast, together with other local resources such as ivory, was important in this trade.

Often, the desire to control the metal-producing regions was a source of conflict and cleavages in society, for example internecine wars were fought to control the gold fields in Northern Zimbabwe (Mudenge 1988). A good example is that of the Almoravids who laid siege at Koumbi Saleh, the capital of the Soninke state of ancient Ghana, thereby precipitating its demise (Levtzion 1973). An ability to control the trade routes and gold producing regions empowered Songhai to suffocate its predecessor Mali. In Southern Africa, the Portuguese desire for gold, fuelled by uncontrolled rapacity, plunged them deep into the internal affairs of the Mutapa state (AD 1450–1900) in Northern Zimbabwe which gradually weakened with the unintended result of nearly halting gold production (Pikirayi 2001). Local agents were not idle while the Portuguese were making their bid for fortune; in the late 17th century, Changamire Dombo fought them and dislodged them from the Zimbabwe plateau, leading to the demise of old and rise of new empires. The lure of African metals also promoted the settlement on African shores by people from its trading partners. In the late 19th century, it was partly as a result of Africa's metal wealth that the continent was colonized by European powers expanding under industrial capitalism. The colonial elites continued to work the mines historically worked by preindustrial people and introduced their own methods of production. This drove the millennia old African techniques into extinction.

One of the most astounding observations is that although Africa was an important cog in the development of "proto" forms of globalization, it did not participate much in the technology transfer taking place between some of its trading partners. According to Killick and Fenn (2012), African metals such as gold paid for exotic commodities in India and Persia as well as the porcelain and gunpowder from China. Yes, Africa obtained commodities from these areas but the incorporation of imported goods was not matched by a technological transfer. For example, the technology of metal production remained the bloomery process, while neither the blast furnace nor technologies of producing *wootz* steel or Damascus swords were adopted in the continent. The same applies to West Africa, which was also in direct contact with the Islamic world. The technology of producing glass was also not developed locally with the exception of the melting of imported glass beads to produce garden rollers at K2 (AD 1000–1200) in the Middle Limpopo valley of Southern Africa and the autochthonous Yoruba glass production industry of the early second-millennium AD in Southwestern Nigeria (see Lankton et al. 2006). It is not that Africans were incapable of assimilating these technologies; rather, there existed deep seated cultural barriers that may have prevented the assimilation of exotic technologies. Furthermore, Hopkins (1973) argues that external methods of production before the industrialization did not have much of an advantage when compared to African ones. For example, David Livingstone commented that African iron was of a better quality when compared to that produced in Europe in the late 19th century (Chirikure 2006). Bloomery iron production was still used in the USA up to the 19th century. The only reason why colonialism managed to change African technologies was that it introduced a different value system based on capitalism and Christianity while directing heavy assaults on local technological practices. Therefore, Africans had no choice, whereas in the past they exercised their freedom by sticking to what worked for them. As such, for as long as the technologies worked, there was no need to change them in favour of alien ones. Also, some of the technologies from Africa's trading partners were suited for contexts with comparatively large and concentrated populations and may not have worked in the variably populated parts of the continent. For example, regions such as Southern Africa were not as heavily populated as the Upper Nile or the Indian subcontinent with the result that large-scale production methods were unlikely to achieve a similar effect.

Conclusion

In summary, it is clear that metallurgy had a significant impact on not just African communities but also those in Eurasia. Far from being isolated, Africa was part of the developments in the Old World. At this point, it is important to revisit the question, did the adoption of metallurgy represent a revolution or not? This is not an easy question to answer because inherently it implicates different scales of analyses, different scales of sociopolitical organization and different scales of population dynamics. As with other places in the world, the adoption of metallurgy was gradual with some communities sticking to the tried and tested technology of stone. However, once metal became more and more incorporated into local value systems, it became so intensely embedded in society that its impact was widely felt.

Metallurgy affected various communities that practised it—with strong consequences for food production, defence, wealth accumulation and local, regional as well as long-distance interconnections. Although metals were known in the western hemisphere, it was only after Columbus that metals became widely used for tool making. However, in areas of the Old World such as Eurasia and Africa, metallurgy played an important part in regional and international integration, promoting urbanization and marked social differentiation. Because of these networks, Africa has always been at the centre of developments in the world. It, however, seems that the continent's population may have been somewhat at a disadvantage because technologies in use in other regions were best suited for sustaining large populations and not small ones. This means that technologies in use on the African continent had to fit this specific context.

While control over metal production and use was a feature of Egyptian, Nubian and Aksumite states, to the extent that stock taking was adhered to religiously, it is difficult to comprehend how rulers in other parts of Africa would have controlled production and metal use. As Mudenge (1974) expressed it, it seems unlikely that there was a rigorous bureaucratic control in the Mutapa and Rozvi states of second-millennium AD Southern Africa. Equally, Levtzion (1973) makes it clear that neither the rulers of ancient Ghana nor Mali controlled the actual production of gold. In the Mutapa and Rozvi states, the commoners based in the villages worked gold, exchanged it for glass beads and cloth, some of which they gave their rulers as tribute.

At a broad scale, the gradual dominance of metallurgy in every aspect of society from the aesthetic, utilitarian, sensual, ceremonial—in fact the whole value system—indicates that metallurgy had significant consequences for social life. If we emphasize continuity, we realize that precious metals such as gold were valuable to Hebrew King Solomon 3000 years ago just as they were to Cecil John Rhodes in the late 19th century. Both sought valuable metals in lands afar showing incidences of history repeating itself but in markedly different contexts. This provides motivation for more work into understanding preindustrial metal production and use. As may be divined from this discussion, it is difficult to understand the beginning and functioning of the ancient and modern worlds without studying metals—a critical part of our sociological and technological past.

References

Adu-Boahen, K. (2006). Pawn of contesting imperialists: Nkoransa in the Anglo-Asante rivalry in Northwestern Ghana, 1874–1900. *Journal of Philosophy and Culture, 3*(2), 55–85.
Anquandah, J. (1993). The Kintampo complex: a case study of early sedentisim and food production in sub-Sahelian West Africa. In T. Shaw, P. Sinclair, B. Andah, & A. Okpoko (Eds.), *The archaeology of Africa food metals and towns* (pp. 255–260). London: Routledge.
Bisson, M. S. (1982). Trade and tribute. Archaeological evidence for the origin of states in South Central Africa (Commerce et tribute. Documents archéologiques sur l'origine des États du sud de l'Afrique centrale). *Cahiers d'Etudes africaines, 22*(87/88), 343–361.
Bisson, M. (2000). Precolonial copper metallurgy: sociopolitical context. In M. Bisson, P. de Barros, T. C. Childs, & A. F. C. Holl (Eds.), *Ancient African metallurgy: the sociocultural context* (pp. 83–146). Walnut Creek: AltaMira Press.
Bocoum, H. (2006). Histoire technique et sociale de la métallurgie du fer dans la vallée moyenne du Sénégal. Doctorat d'Etat thesis. University Cheikh Anta Diop, Dakar
Bohannan, P. (1955). Some principles of exchange and investment among the Tiv. *American Anthropologist, 57*(1), 60–70.
Carneiro, R. L. (1967). On the relationship between size of population and complexity of social organization. *Southwestern Journal of Anthropology, 23*(3), 234–243.
Childe V. G. (1930). *The bronze age*. Cambridge: Cambridge University Press.

References

Chirikure, S. (2006). New light on Njanja iron working: towards a systematic encounter between ethnohistory and archaeometallurgy. *South African Archaeological Bulletin, 61*, 142–151.

Chirikure, S. (2007). Metals in society: Iron production and its position in Iron Age communities of Southern Africa. *Journal of Social Archaeology, 7*(1), 72–100.

Chirikure, S., Heimann, R. B., & Killick, D. (2010). The technology of tin smelting in the Rooiberg Valley, Limpopo Province, South Africa, ca. 1650–1850 CE. *Journal of Archaeological Science, 37*(7), 1656–1669.

Chirikure, S., Manyanga, M., Pikirayi, I., & Pollard, M. (2013). New pathways of sociopolitical complexity in Southern Africa. *African Archaeological Review, 30*(4), 339–366.

Cline, W. B. (1937). *Mining and metallurgy in Negro Africa* (No. 5). George Banta Publishing Company

Connah G (2001). *African civilization*. 2nd edition. Cambridge: Cambridge University Press. (Conservation Institute; J. Paul Getty Museum, in association with Archtype Books).

David, N., & Sterner, J. (1997). Water and iron: phases in the history of Sukur. In H. Jungraithmayr, D. Barreteau, & U. Seibert (Eds.), *L'homme et l'eau dan le basin du lac Tchad* (pp. 255–270). Actes du Colloque Mega-Tchad. Paris: Colloques et Seminaires, Editions ORSTOM.

de Barros, P. (1986). Bassar: A quantified, chronologically controlled, regional approach to a traditional iron production centre in West Africa. *Africa, 56*(2), 152–174.

de Maret, P. (1985). The smith's myth and the origin of leadership in Central Africa. In R. Haaland & P. Shinnie (Eds.), *African iron working—ancient and traditional* (pp. 73–87). Oslo: Norwegian University Press.

Fagan, B. M. (1969). Early trade and raw materials in South Central Africa. *Journal of African History, 10*(1), 1–13.

Fletcher, R. J. (1993). Settlement area and communication in African towns and cities. In T. Shaw, P. Sinclair, B. Andah, & A. Okpoko (Eds.), *The archaeology of Africa food metals and towns* (pp. 732–749). London: Routledge.

Garlake, P. S. (1968). The value of imported ceramics in the dating and interpretation of the Rhodesian Iron Age. *The Journal of African History, 9*(1), 13–33.

Garlake, P. S. (1973). *Great Zimbabwe*. London: Thames and Hudson.

Garrard, T.F. (1980). *Akan weights and the gold trade*. London: Longman.

Garrard, T. F. (2011) (1989). *African Gold: jewellery and ornaments from Ghana, Côte d'Ivoire, Mali andSenegal in the collection of the Barbier-Mueller Museum*. Munich: Prestel.

Grant, M. R. (1999). The sourcing of Southern African tin artefacts. *Journal of Archaeological Science, 26*(8), 1111–1117.

Hall, S. L. (1998). A consideration of gender relations in the Late Iron Age "Sotho" sequence of the Western Highveld, South Africa. In S. Kent (Ed.), *Gender in African prehistory* (pp. 235–260). Walnut Creek: AltaMira Press.

Hall, S. L., Miller, D., Anderson, M., & Boeyens, J. (2006). An exploratory study of copper and iron production at Marothodi, an early 19th century Tswana town, Rustenberg District, South Africa. *Journal of African Archaeology, 4*, 3–35.

Hauptmann A. (2007). *The archaeometallurgy ofc Copper: Evidence from Faynan, Jordan*. New York: Springer.

Herbert, E. W. (1984). *Red gold of Africa: copper in precolonial history and culture*. Madison: University of Wisconsin Press.

Holl, A. F. (2009). Early West African metallurgies: new data and old orthodoxy. *Journal of World Prehistory, 22*(4), 415–438.

Hopkins, A. G. (1973). *An economic history of West Africa*. New York: Columbia University Press.

Horton, M. C., & Middleton, J. (2000). *The Swahili: the social landscape of a mercantile society*. Oxford: Blackwell.

Humpris, J., & Iles, L. (2013). Pre-colonial iron production in Great Lakes Africa: recent research at UCL Institute of Archaeology. In J. Humpris & Th. Rehren (Eds.), *The world of iron* (pp. 56–64). London: Archetype.

Huysecom E., & Agustoni B. (1997). *Inagina: l'ultime maison du fer/the last house of iron*. Geneva: Telev. Suisse Romande. (Videocassette, 54 min).

Killick, D. J. (2009). Agency, dependency and long-distance trade: East Africa and the Islamic World, ca. 700–1500 C.E. In S. Falconer & C. Redman (Eds.), *Polities and power: archaeological perspectives on the landscapes of early states* (pp. 179–207). Tucson: University of Arizona Press.

Killick, D., & Fenn, T. (2012). Archaeometallurgy: The study of preindustrial mining and metallurgy. *Annual Review of Anthropology, 41*, 559–575.

Killick, D. (2014). Cairo to Cape: the spread of metallurgy through Eastern and Southern Africa. In B. W. Roberts & C. Thornton (Eds.), *Archaeometallurgy in global perspective* (pp. 507–527). New York: Springer.

Kim, N. C., & Kusimba, C. M. (2008). Pathways to social complexity and state formation in the Southern Zambezian region. *African Archaeological Review, 25*(3–4), 131–152.

Kusimba, C. M. (1999). *The rise and fall of Swahili states*. Walnut Creek: AltaMira Press.

Lankton, J., Ige, A., & Rehren, T. (2006). Early primary glass production in Southern Nigeria. *Journal of African Archaeology, 4*, 111–138.

LaViolette A., & Fleisher J. (2005) The archaeology of sub-Saharan urbanism: Cities and their countrysides. In A. B. Stahl (Ed.), *African Archaeology: A critical Introduction* (pp. 327–352). Oxford: Blackwell.

Levtzion, N. (1973). *Ancient Ghana and Mali* (Vol. 7). London: Methuen.

MacDonald, K. C. (1998). Before the empire of Ghana: Pastoralism and the origins of cultural complexity in the Sahel. In G. Connah (Ed.), *Transformations in Africa: Essays on Africa's later past* (pp. 71–103). London: Leicester University Press.

MacEachern, S. (1993). Selling the iron for their shackles: Wandala-Montagnard interactionsin Northern Cameroon. *Journal of African History, 34*, 247–247.

Mackenzie, J. M. (1975). A pre-colonial industry: the Njanja and the iron trade. *Nada, 11*(2), 200–220.

Maggs, T. (1992). 'My father's hammer never ceased its song day and night': the Zulu ferrous metalworking industry. *Southern African Humanities, 4*, 65–87.

Magnavita, C., Breunig, P., Amfje, J., & Posselt, M. (2006). Zilum: a mid-first millennium BC fortified settlement near lake Chad. *Journal of African Archaeology, 4*, 153–169.

McIntosh, S. K. (Ed.). (1999). *Beyond chiefdoms: pathways to complexity in Africa*. Cambridge: Cambridge University Press.

McNaughton, P. R. (1993). *The Mande Blacksmiths: knowledge, power, and art in West Africa*. Bloomington: Indiana University Press.

Miller, D. (1996). *Tsodilo Jewellery*. Cape Town: University of Cape Town Press.

Miller, D. (2001). Metal assemblages from Greefswald areas K2, Mapungubwe Hill and Mapungubwe Southern terrace. *The South African Archaeological Bulletin, 56*, 83–103.

Mitchell, P. (2002). *The archaeology of Southern Africa*. Cambridge: Cambridge University Press.

Mitchell, P., & Lane, P. (Eds.). (2013). *The Oxford handbook of African archaeology*. Oxford: Oxford University Press.

Mudenge, S. I. (1974). The role of foreign trade in the Rozvi empire: A reappraisal. *Journal of African History, 15*(3), 373–391.

Mudenge, S. I. (1988). *A political history of Munhumutapa c 1400–1902*. Harare: Zimbabwe Publishing House.

Ndoro, W. (1994). Natural draught furnaces south of the Zambezi River. *Zimbabwean Prehistory, 14*, 29–32.

Nixon, S. (2009). Excavating Essouk-Tadmakka (Mali): new archaeological investigations of early Islamic trans-Saharan trade. *Azania: Archaeological Research in Africa, 44*(2), 217–255.

Nixon, S., Rehren, T., & Guerra, M. F. (2011). New light on the early Islamic West African gold trade: coin molds from Tadmekka, Mali. *Antiquity, 85*(330), 1353–1368.

Ogundiran A (2005). Four millennia of cultural history in Nigeria (ca. 2000BC–AD 1900). *Journal of World Prehistory, 19*(2), 133–168.

Phillipson, D. W. (2005). *African archaeology*. Cambridge: Cambridge University Press.

Phimister, I. R. (1974). Alluvial gold mining and trade in nineteenth-century South Central Africa. *The Journal of African History, 15*(3), 445–456.

References

Pikirayi, I. (2001). *The Zimbabwe culture: origins and decline in Southern Zambezian states*. Walnut Creek: Altamira Press.

Prestholdt, J. (2008). *Domesticating the world: African consumerism and the genealogies of globalization*. Berkeley: University of California Press.

Pwiti, G. (1996). *Continuity and change: an archaeological study of farming communities in Northern Zimbabwe AD 500–1700*. Uppsala: Societa Archaeologica Uppsaliensis.

Pwiti, G. (2005). Southern Africa and the East African coast. In Stahl, A. B. (Ed.), *African archaeology: a critical introduction* (pp. 378–391). London: Blackwell.

Reid, A., & MacLean, R. (1995). Symbolism and the social contexts of iron production in Karagwe. *World Archaeology, 27*(1), 144–161.

Renfrew, C., & Cherry, J. F. (Eds.). (1986). *Peer polity interaction and socio political change*. Cambridge: Cambridge University Press.

Ringquist. J. (2008). Kongo iron: Symbolic power, superior technology and slave wisdom. African diaspora archaeology network. *Newsletter* (Sept):1–20.

Robertshaw, P., Wood, M., Melchiorre, E., Popelka-Filcoff, R. S., & Glascock, M. D. (2010). Southern African glass beads: chemistry, glass sources and patterns of trade. *Journal of Archaeological Science, 37*(8), 1898–1912.

Sasson, H. (1964). Iron-Smelting in the Hill Village of Sukur, North-Eastern Nigeria. *Man, 64*, 174–178.

Scott, D. A. (1991). *Metallography and microstructure of ancient and historic metals*. Getty Conservation Institute; J. Paul Getty Museum, in association with Archtype Books.

Shaw, T. (1970). *Igbo-Ukwu: an account of archaeological discoveries in Eastern Nigeria* (Vol. 2). Evanston: Northwestern University Press.

Smith, C. S. (1981). *A search for structure. Selected essays on science, art and history*. Cambridge: MIT.

Stanley, G. H. (1931). Some products of native iron smelting. *South African Journal of Science, 28*, 131–134.

Stayt, H. A. (1931). *The Bavenda* (No. 58). International Institute of African Languages & Cultures. Oxford: Oxford University Press.

Summers, R. (1969). *Ancient mining in Rhodesia and adjacent areas*. Salisbury: Trustees of the National Museums of Rhodesia.

Tamari, T. (1991). The development of caste systems in West Africa. *Journal of African traditional iron production centre in West Africa. Africa, 56*(2), 152–174.

Vansina, J. (1969). The bells of kings. *The Journal of African History, 10*(02), 187–197.

Warnier, J. P., & Fowler, I. (1979). A nineteenth-century Ruhr in Central Africa. *Africa, 49*(04), 329–351.

Wood, M. (2012). *Interconnections: glass beads and trade in Southern and Eastern Africa and the Indian Ocean-7th to 16th centuries AD*. African and Comparative Archaeology. Uppsala: (Department of Archaeology and Ancient History), Uppsala University.

Chapter 7
Bridging Conceptual Boundaries, A Global Perspective

> *"Pre-European metalworkers are worthy of respect for the results they achieved with primitive methods"*
> (Steel 1975, p. 232)

Introduction

The production and use of metals in antiquity rank as one of the humanity's most consequential sociocultural and technological developments, which make the study of Africa's preindustrial metallurgy a topic of global significance. Although Africa is part of the Old World, its pathway to metallurgy was remarkably different to that of its counterparts. Unlike regions such as Thailand which are believed based on current evidence to have received metallurgy from China in exactly the same succession of copper, bronze and iron (Pryce et al. 2010), it appears as if some parts of Africa adopted metallurgy in reverse of the situation throughout Eurasia. While Egypt, Nubia, Ethiopia and Eritrea followed the Eurasian trajectory, most of West, Central, Southern and Eastern Africa started metallurgy with iron and in some instances iron and copper and only adopted bronze, gold and tin more than a thousand years later. Not surprisingly, researchers are not agreed on the source of Africa's metallurgy. Proponents of the external origins hypothesis argue that given the importance of early encounters with pyrotechnology and the radiocarbon black hole between 800 and 400 BC, it seems that knowledge of African metallurgy diffused from the Middle East (Alpern 2005). However, those who support local origins maintain that vast differences in pathways between the Middle East and Africa, and increasing numbers of dates clustered around 1000 BC indicate that African metallurgy may be independent (Holl 2009). The acceptability of the two positions is affected by poor radiocarbon dates, poor contexts of recovery and too few sites that have been excavated, as well as the challenges created by calibration in the critical period between 800 and 400BC. The technical reasons supporting the external origins hypothesis, such as the difficulty of smelting iron before mastering easier metals, make considerable scientific sense (Alpern 2005; Clist 2013).

However, the big question is, given that Eurasia had knowledge of the materials and materiality of iron, lead, copper, bronze and gold, during the time when African metallurgy began, why did Africans only choose to take up iron and copper and not the other metals and alloys? Although it is possible that people without a prior knowledge of the materials and materiality of metals were able to exercise choices informed by the physical properties of metals to the extent of selecting the more durable iron, it is not clear why they would ignore gold which had a very high value and was highly sought after. If materiality was the biggest motivator for the adoption of early metallurgy (Smith 1981), then iron is certainly not the most beautiful of all metals, implicating other reasons. Furthermore, why did sub-Saharan Africa accept tin, bronze and gold almost a thousand years later? It has been shown that the adoption of technology is a result of prior experiences (Prestholdt 2008). As such, it was easier to accept gold once copper and iron were known. If Craddock (2010) is right, then sub-Saharan iron may have evolved out of Nubian metallurgy.

However, chance may have promoted the advent of iron metallurgy; temperatures for smelting iron and copper do not vary much, given that gangue minerals such as silica in the ore were fluxed to form slags at temperatures over 1000 °C (Bachmann 1982). The dynamics of innovation, diffusion and technology and value transfer are neither simple nor logical and may be easily discerned through common sense rooted to a specific cultural logic. In other words, we should endeavour to understand these transfers in relation to local, culturally specific logics. It has been argued that iron smelting is complicated, but there is skepticism that people without experience with pyrotechnology can get an idea from somewhere and execute that difficult idea with relative ease. Thus, if iron smelting is as difficult as some believe, then it is unlikely that a mere transmission of ideas would result in a transfer that we see in West, Central and East Africa. As Holl (2009) has argued, there is no hard evidence indicating technology transfer between sub-Saharan Africa and the donor regions. The copper at Akjoujt may represent a different dynamic that indicates what happens in zones of contact. Had this happened in other areas, the idea of diffusion would make sense; however, Akjoujt seems to be an exception rather than the rule. Because some of the greatest inventions in human history were mere accidents, what is required is more fieldwork backed up by robust assessment of contexts of recovery and dating programs so that the conclusions are based on hard evidence. The significance of the great controversy regarding the origins of African metallurgy is that it forces researchers to reflect on innovation as well as value and technology dispersal in extremely varied contexts.

By the mid-first millennium AD, sub-Saharan Africa was fully integrated into the triangular trading system involving Africa, the Near East and South Asia. In addition to other commodities, Africa supplied gold, iron, and possibly copper and tin, which were in demand within Islamic-controlled Saharan and Indian Ocean trade networks. In return, Eurasia supplied glass beads, brass, bronze, silver and among other resources cloth and later ceramics, augmented by small quantities of celadon and porcelain. African metallurgy was, therefore, instrumental in the development of land-based as well as sea-based linkages and therefore connections internal to the continent and externally with Eurasia. This interaction resulted in mobility and settlement in Africa by Arabs along the East African littoral and north of the Sahara. The Chinese under the legendary Captain Zheng Ho anchored on the East African coast once 60 years before Vasco Da Gama. Over time, the Europeans followed suit and built a number of castles and forts alongside the Atlantic and Indian Ocean coasts. Such activities resulted in the inward and outward exchange of ideas, diseases, animals and metals. However, strong cultural filters enabled Africa to only incorporate commodities and practices that aligned with extant values and systems of valuation, while ignoring those unsuited in its context. Also, simple exposure to new practices or ideas does not necessarily lead to adoption of those practices or ideas. There is little evidence in much of Africa of assimilation of either Western or Asian architecture and methods of metalworking, despite the many levels of interaction.

This culture contact and interaction contradicts views that have traditionally profiled Africa as a cultural backwater isolated from the rest of the world (Stahl 2014a; Mitchell 2005). Neither was Africa passive nor was it at the mercy of its trading partners. Instead, it exercised a great deal of agency, choice and initiative that resulted in the continent co-opting what worked in its context while rejecting that which did not suit. Rather than being at the mercy of other regions, an encounter with African metallurgy indicates that Africans in large part controlled their destiny. Therefore, it is important to put aside the ideas relating to both overwhelmed Africa and overwhelming Eurasia as well as backward Africa and advancing Eurasia because different dynamics involving population sizes and cultural contexts were at play.

Besides interaction and culture contact, African metallurgy is rich in symbolic and cultural information regarding the different stages in the *chaîne opératoire* of metal production and use. The success of African metal working was dependent on technical skill as much as on the associated belief systems, an observation that equally applies to most parts of the world with knowledge of metallurgy. There is a great deal of versatility and diversity in furnace types, the range and quality of ores worked and the methods of provisioning air to the furnaces.

African Metallurgy and the Bridging of Conceptual Boundaries Between Technology, Society and Culture

Any academic excursion into Africa's preindustrial metallurgy shows that its success was based on the intense interaction between nature and culture to the extent that the boundary between the two became increasingly blurred as nature was transformed into culture. For example, the raw materials for metallurgy—ore, clay and wood—conceptually came from nature, such that metalworkers appealed to the power of ancestors, deities and the supernatural. These raw materials were transformed into cultural products through smelting and the subsequent smithing and fabrication. The metal, representing a product from nature, became not just a cultural commodity in the service of society but also a medium through which culture interacted with nature. Ethnographies conducted from the late 19th century onwards demonstrate the intertwined nature of technology, science and ideology in the African past. Therefore, a fusion of magic, science, society and belief systems is an important element of African technological practice, repertoire and style just as they were part of medieval European science and technology (Hansen 1986). As such, global archaeology will only be poorer if it ignores the web of entangled and associated beliefs that were integral to the success of metallurgy. Insoll (2008) has demonstrated the utility of this approach with respect to ritual practice and shrines, using insights from Northern Ghana to generate new perspectives on European Neolithic and Bronze Age contexts.

In this regard, African metallurgy provides one of the best examples of pre-capitalist technologies of practice and technological style. For example, historical documents, ethnoarchaeological studies and archaeological field research show that in many African societies, the smelting of metal was metaphorically equivalent to copulation, gestation and birth. The furnace was metaphorically perceived to be a woman, while smelting was analogous to intercourse and gestation. The metal bloom growing in the furnace was akin to a fetus, while male smelters were simultaneously husbands and midwives (Schmidt 2009; Killick and Fenn 2012). In some instances, this symbolism was written on the furnaces as evidenced by decorations of female breasts, genitalia, navels and waist belts worn to enhance fertility (Childs 1991). The reduced metal was conceptually seen as a child, who passed through several stages of life and in turn contributed to fertility of the land and society in general. Not surprisingly, in some regions, iron hoes could be exchanged for women's reproductive power resulting in societal renewal and growth.

Bridging Analytical Boundaries: From Sources of Ethnographies to Domains of Integrated Studies

Whether one considers Africa as an independent inventor of metallurgy or a receiver from the Middle East and adjacent territories does not matter much. Rather, the most germane topic that also envelopes elements of the origins issue is what is the value of African preindustrial metallurgy in understanding global archaeology. As I have argued elsewhere (Chirikure 2010), there are two Africas in global archaeology: The first is the origin of humanity and centre of cultural developments for approximately 95 % of human history; the second is a more recent Africa, perceived as backward in sociocultural developments from the mid-Holocene onward. As such, it is a continent whose significance in knowledge production on Pleistocene human development is acknowledged globally on the one hand, and a continent perceived as having little to offer in the study of the last 2000 years, except perhaps as a source of ethnographies for validating models developed elsewhere. A question that arises is what motivated these contrasting academic positions? Obviously, this implicates the context of knowledge production and how research topics are articulated locally and globally.

According to Robertshaw (1990), from the late 19th century, Africa was often depicted as a backward continent that still hosted metallurgical practices long extinct elsewhere. Therefore, the continent's recent past became a rich source of analogues for developing interpretations in other places. According to Kense and Okoro (1993), this "Africa as source of ethnographies" paradigm dictated that the continent was not seen as a region possessing technological developments with potential to provide different, if not independent trajectories of the evolution of metallurgy. Rather, it was the source of analogues for interpreting early metallurgy in other continents using its ethnographic record. And yet, Africa's preindustrial metallurgy metamorphosed in different directions, which warrant detailed comparisons with practices elsewhere to develop a global picture on technological variation and cross-borrowing (Chirikure 2005).

At different intervals in the 20th century, a number of smelting re-enactments were carried out in Africa to record the process before the knowledgeable practitioners passed on. Because such work was carried out by men and women who understood and appreciated the relevance of African preindustrial metallurgy, it was mostly comprehensive in its approach. Most of the time, the recording started with raw material collection, through smelting and smithing to the final products and waste materials. Schmidt (1978, 1997), for example, as discussed in Chap. 4, combined oral histories, ethnographies and the archaeology and archaeometallurgy to develop a long-term perspective on iron production in Tanzania. While some of Schmidt's insights, for example, the pre-heating hypothesis discussed above, have been challenged, his work was important for attempting to tease out important innovations relating to African smelting. Overall, Schmidt's integrated work demonstrated that it is important to engage with early history of African metallurgy to understand its development and evolution. This is important because what seemed like a "fossil" technology in the early 20th century was vibrant, dynamic and innovative across the ages as human beings responded to various challenges, technological or otherwise. This underscores the value of comparative approaches conducted across long time scales in identifying the problematic character of projecting ethnographic practices into the past. For example, the variability of spatial location of smelting inside recent historical settlements and those of the Early Iron Age villages in Southern and Eastern Africa indicates the ability of long-term perspectives in teasing out continuities and changes.

Although Cline (1937) sketched much in terms of ethnographic distribution of techniques of metalworking, rarely did researchers study the different varieties of metalworking in their context to elicit their most salient features. David et al. (1989) combined ethnographic and archaeometallurgical study of Mafa smelting in Cameroon discussed in Chap. 4 and revealed a technology between bloomery and blast furnace for it could produce soft and cast iron. Within Africa, these in-between processes contrast remarkably with those that produced usable iron from very low grade ores and the comparatively smaller bowl furnaces that equipped Shaka's armies discussed in Chap. 5. The mechanisms of operating furnaces varied from slag-tapping bowl furnaces to non-slag-tapping natural draught furnaces and from slag-tapping natural draught furnaces to non-slag-tapping low shaft and bowl furnaces. The bellows too differed from area to area but in a complex mix. This prompted Schmidt (2001) to argue that referring generically to African metallurgical processes as bloomery techniques obscures this anourishing variation and diversity. Therefore, Africa offers an unrivalled potential to foster understanding not just of variability in technological practice but also diversity in the materials and materialities of preindustrial metallurgy. There is in fact a much wider variety of bloomery iron smelting processes documented in Africa than elsewhere. Is it possible that these were also once practiced elsewhere in Eurasia but were erased by the spread of the blast furnace and finery? On the other hand, African copper smelting technology seems very restricted compared to other parts of the world. Why was this? Perhaps because copper smelting was never a major technology within the continent and that Eurasia had a very long time to experiment with copper from c. 5000 BC to c. 1500. Had Africa started with copper, perhaps it too would exhibit such diversity.

Global archaeology is quite creative when it comes to developing new theories. For example, given the bifurcation of studies of preindustrial metallurgy into the scientific and magical, researchers such as Lechtman (1977) propositioned, following Mauss, that all technologies were socially constructed and embedded. The different techno–cultural solutions gestated in different parts of the world shaped diverse technological styles—distinctive ways of doing things that in most cases achieve more or less the same result. By the 1990s and 2000s, the concept of *chaîne opératoire*, borrowed from French anthropology, sought to explore the different elements of these technological styles. Within that paradigm, scholars seek to simultaneously consider the physical properties and cultural ethos associated with technologies. While this is ground-breaking work in global archaeology, Africanists have long recognized and appreciated Africa's divergent technological styles (see, for example, Rickard 1939; Cline 1937). Therefore, the view that technology is culturally embedded and is socially constructed has always been axiomatic in studies of Africa's preindustrial metallurgy—if only global archaeology was paying more attention!

Given that it is now universally acknowledged that Africa's preindustrial technologies were culturally mediated, responses to a specific situation and not relics of technologies extinct elsewhere, it is important to invest more in understanding the process from an interdisciplinary and diachronic point of view. As an example, a study by Heimann et al. (2010) focusing on the complex mineralogy of tin smelting slags from Southern Africa has greatly enhanced our understanding of the intricacies of preindustrial tin smelting. Furthermore, an exploratory statistical study of slags from African archaeological sites conducted by Chirikure and Bandama (2014) indicated that different furnaces used in preindustrial Africa had different reduction capacities. The tall natural draught furnaces were more reducing, such that their slag composition appeared closer to equilibrium than those from low shaft and bowl furnaces when plotted on ternary diagrams. The decision to choose a specific furnace type depended on many other variables such as efficiency in labour and time, quality of ore and nature of smelting technology (Chirikure and Rehren 2006). Clearly, there is more to be learned from Africa regarding technological development beyond being a mere source of ethnography. At the same time, the ethnoarchaeological/ethnographic insights into the importance of ritual in African technological practice hold potential to enrich studies of metallurgy elsewhere—less in terms of the specifics of ritual practice, and more in relation to the culturally embedded character of what is often perceived as "purely" technological practices (Stahl 2014a, b).

Local Responses to Technology Transfer and Knowledge Dispersal

The subject of Africa's preindustrial metallurgy is intricately interwoven with broader issues related to technology transfer and knowledge transmission between interacting peoples. As part of the Old World, Africa has always explicitly and implicitly interacted with various polities in Eurasia. The resources of the Egyptian Sudan and adjacent margins contributed to the Middle Eastern political economy. Black Nubian monarchs also ruled Egypt as the 25th Dynasty. However, the connections between North Africa and the Middle East on the one hand and sub-Saharan Africa on the other before the onset of the first millennium AD are at best vague and at worst opaque. The presence of this "black hole" makes it difficult to understand cultural and knowledge exchanges between regions to the north and south of the Sahara. Obviously, the origins of African metallurgy features strongly in this debate because the routes for the supposed north to south transmission of knowledge have not been well articulated, not least because of a lack of research and the concomitant deficiencies of radiocarbon in the interval 800–400 BC. Perhaps, the most important issue as far as technological transfer is concerned with respect to origins questions centres on the time lag between adopting a new technology and modifying it to suit the local situation. Surely, the outward differences between the metallurgy of the source areas and that of sub-Saharan Africa suggest a complex and confusing technology transfer mechanism (if such a transfer happened at all).

The issue of innovation is intimately associated with the local origins thinking and the subsequent development of the technology. The lesson from the past is that innovations hardly follow "common sense". Often an accident results in a discovery of huge significance. Yes, it is pyrometallurgically more difficult to smelt iron than copper, but some furnaces used in preindustrial Africa comprised rudimentary structures—in some cases made of banana stems—that still produced iron (Celis and Nzikoyabanka 1976). Furthermore, temperatures for reducing copper are not that different from those for reducing iron given the fact that typical iron and copper ores contained gangue materials which required slag formation at temperatures above 1000 °C. The only answer lies in more research at sites with evidence of early African metallurgy. Fundamentally, some perspectives that threaten acquired knowledge or the orthodoxy may not be widely accepted (Holl 2009). For example, would the Catholic Church of the Middle Ages have envisaged that Copernicus' idea that the world was a globe would become an unquestioned fact? Whatever the mechanism for origins and innovation, once established, it is clear that Africa's preindustrial metallurgy underwent regional and context-specific innovations, which produced a very diverse range of technological styles from the large-scale production in West Africa to the small-scale production in comparatively smaller furnaces in Southern Africa.

That Africa did not adopt metalworking traditions of the Eurasian world after increased contact between these two areas in the first millennium AD underscores that the adoption of technologies is a culturally mediated experience. In addition, some of the Eurasian technologies were designed for very large populations, which were not an issue in sparsely occupied areas of sub-Saharan Africa. That existing technologies could meet local demand was all that mattered. Even so the quality was admirable and in some cases better than that produced in the so-called advanced blast furnaces of Europe in the mid-19th century (Chirikure 2006).

African Metals: Land and Sea Links and Protoforms of Globalization

Africa has always been part of the Old World, participating in different social and economic relations. The varying mineral availability gradients precipitated contact between different areas. Internally in Africa, this connected different communities. However, it seems that technological styles were an important aspect of identity, for it was rare for Africans to adopt methods of the others. Different regions of Africa such as Southern and Central Africa were well networked, just as those of West and North Africa. In Southern Africa too, contact and cultural exchanges took place between the coastal communities and hinterland communities. These land links were vital for long-distance trade for the objects that formed the life blood of this system came from hinterland areas.

The long-distance trade that linked Southern Africa and the Indian Ocean world via coastal East African communities resulted in a long-distance network spanning continents. African gold, iron and other resources such as ivory were exchanged for Indo-Pacific and Persian glass beads, Persian ceramics and Chinese porcelain. Eurasian alloys such as brass, leaded gun metal and bronze were also brought to Africa. Despite this contact, technology transfer was minimal. The trans-Saharan trade also introduced trade beads and brass to West Africa just as the Atlantic trade based in the same region. The commodities from long-distance trade were incorporated into local value systems where their possession was a source of political leverage. Those who controlled the trade, sometimes had access to political power. However, in some cases, the trade was very open with citizens free to participate in the trade and the elites were not always in control. Political power was based on control over land and ideology and not distribution and redistribution of imports (Bhila 1982).

In this contact, not all objects and technologies were locally accepted in Africa. Brass was imported, but worked using local techniques. This shows that the acceptance of technologies is culturally specific. Prestholdt (2004) notes that metal manufacturers in the USA adjusted their standards of copper wire to meet the demands of East African trade, producing gauge acceptable to groups such as Maasai. Similarly, in their encounter with Southern Africa, the Portuguese were frustrated because its inhabitants shunned European glass beads, preferring those from India. Africa supplied commodities that through trade networks enabled people in Eurasia to have access to luxuries and necessities such as porcelains and gun powder among other commodities. Therefore, the world has always depended on the mother continent for raw materials. Underdevelopment theorists such as Rodney (1974) argue that this position disadvantaged the continent, but it is also a result of the disruption of the African value system at conquest.

Changing Contexts of Knowledge Production and the Future of African Preindustrial Metallurgy

> I would remark that of the woolly haired Africans, who constitute the principal part of the inhabitants of Africa, there is no history, & there can be none. That race has remained in barbarism from the first ages of the world; their country has never been explored very fully by civilized man (Webster cited by Yacovone 2002, p. ii)

According to Holl (2000, p. 6) "throughout the colonial period, sub-Saharan Africa was considered a backward continent on the receiving end of technological innovations". For example, preindustrial metal production in the subcontinent has historically been viewed as derivative in its origins and retarded in its development (Rickard 1939). There was a popular belief in the West that claimed that human societies had evolved through several stages from savagery through barbarism to civilization (see Robertshaw 1990). Whereas Europeans perceived themselves as having reached the civilized stage with their level of technological sophistication, Africa was perceived to be still languishing at the foot of the development ladder in savagery. It is, therefore, not surprising that explorers, travelers and colonialists who either spread throughout the continent or commented on Africa on the eve of colonization argued that Africans occupied a stage in the evolutionary tree that Europeans had passed a thousand or more years ago (Hall 1987, p. 5).

In 1843, a black American religious leader and anti-slavery campaigner resident in New Haven, Connecticut, Amos G. Beman approached Noah Webster, the famed propagator of the first American Dictionary for guidance on what were the best texts about the African people. According to Yacovone (2002, p. ii), Webster responded with the quote that opens this section. However, if Mr. Beman was interested in northern fringes of the continent such as Egypt, Webster retorted that any good encyclopedia was good enough. The net result of such notions was the nurturing of long-lasting stereotypes fuelled by biased social values and Western racial priorities (Yacovone 2002). Going back to Hegel in the 1820s, African societies and technologies such as iron working were thought to be in a "deep and perpetual slumber" without any advancement (Brown 1973, p. 3; Curtin et al. 1978; Goody 1971). On his part, Stanley (1931) argued that "from about 1200 BC onward to the making of iron in the present by the Negroes, the production of iron has been entirely the same that I know no way of distinguishing it..." (Stanley cited in Caton-Thompson 1931, p. 201).

Ritual and symbolism or the materiality of African metal production was not spared from derogatory and biased perceptions. For example, when he failed to understand the significance of symbolic dimensions of Venda iron workers in South Africa, Beuster (1889, cited in Rickard 1939, p. 89) posited that, "it was the ancient custom ... for the smith to add human flesh to the ore in order that iron might make a good hoe, and if no flesh was available the smith sought for it among the dead". As an expert on Venda ethnography, Van Warmelo (1935) did not find any evidence for scavenging the dead, and thus, such statements resonate very well with notions of a dark and savage Africa. Of course, Plug and Pistorius (1999) report the recovery of human finger bones from pits in the floor of some furnaces at Phalaborwa, but it is unlikely that they were scavenged from the dead who are viewed as sacred in the area. It would, therefore, appear that the whole idea was exaggerated by Beuster, in so doing fulfilling stereotypical views of African society, culture and technology.

However, as we have seen, if the multiple furnace types and technological styles that dot the archaeological landscape across the African continent are considered, there is ample evidence for considerable historical and regional diversity, innovation and variation (Cline 1937; Miller et al. 2001; Okafor 1993; Prendergast 1975; Schmidt 1997; Sutton 1985). The Mafa smelters of Northern Cameroon, for instance, produced cast iron from their furnaces, a product otherwise restricted to the blast furnace method introduced after colonization (David et al. 1989). The Fipa of Tanzania employed both bellows-driven and natural draught furnaces. Okafor (1993) has reported on the existence of slag tapping in the Late Iron Age of Nsukka, eastern Nigeria, a technological development not documented in the preceding Early Iron Age of the region. The Njanja devised ways of increasing the air to fuel to ore ratio and reorganized their production by employing a shift system of labour to initiate a hugely successful metal production enterprise (Chirikure 2006). This diachronic and spatial diversity in the technologies of practice in preindustrial metallurgy across sub-Saharan Africa unravels an important source of comparative cases in any study of the different trajectories of metallurgy locally, regionally and globally. For example, why did one type of iron smelting furnace (the tall natural draft furnace) spread so widely in the second millennium AD, but other technologies, like the Mafa furnaces, have not spread beyond the Mandara Mountains? This implicates the presence of historical and demographic processes at work in both cases.

Since its advent, the material and materiality of African metal production played an important role in making items for economic, social and political ends as shown by tools such as hoes, art that represented leaders, and trade and exchange relationships that linked West Africa, North Africa, and the Middle East and Southern Africa and the Indian Ocean rim on the other. Thus, metalworking was a major nexus that fed the heart of local, regional and international interconnections, with the consequence that it stimulated varied economic, political and economic systems spatially and diachronically. Therefore, rather than being, isolated, desolate, and a cultural backwater, Africa lay at the heart of the development of the world from very early on. It is impossible to discuss the success of various generations of Eurasians, without the sterling role that Africa played in supplying raw materials, finished products and ideas in history. The Islamic world system thrived on African gold from the mid- to late first millennium AD onward, and the same applies to the Indian subcontinent. Later in history, African metals paid for European luxuries and in the 19th century was one of the principal reasons for the scramble and partition of Africa. As such, the world's history, values and attitudes are all inscribed in the development of Africa and its interactions with the rest of the world at different temporal scales.

Conclusion

It is clear that metals played an important role in African societies and those of the other continents. There are so many lessons for understanding technology in the world based on the African experience. Africa offers a different technological itinerary as far as metallurgy is concerned. As such, one of the greatest enigmas of our time is the origins of African metallurgy. Although this question is still debated, it has implications for understanding technology transfer and innovation. Innovations are technologically and culturally specific and often result from accidents.

The process of metal production is socially embedded, evoking the power of the ancestors, magic and the deities. It is, therefore, a fallacy of late 19th-century and 20th-century science that magic had no role in science and technology (Hansen 1986). This universal was witnessed in other parts of the world such as Asia and Latin America, showing that metallurgy participated in the production and reproduction of society and was therefore an integral component. The differences in value systems may have prevented technology transfer between various regions. There is need to consider the role of population size in technology transfer and change. Huge populations require technologies suited for servicing them as do small populations, and such differences often motivate variations that are visible archaeologically.

Because metallurgy participated in the production of society, it illuminates interesting gender division of labour and cross-borrowing. The full production chain of metallurgy included a special role and place for women and their material culture. Winnowing was adopted for panning gold in Northern Zimbabwe just as grinding was used to separate ore from waste materials. Most crucibles used in Southern and Central Africa were domestic pots made by women. This is hardly surprising because crucibles, pots and furnaces are all containers important for heat-mediated transformations. Metal production represented men appropriating symbolically the reproductive power of women. As such, while some studies are quick to make distinctions between metallurgy as a purely male domain and other pursuits which were female, the division is not crystal clear as there were so many cross-overs, metaphorically and in practice.

Finally, far from being isolated, Africa played an important role in the development of the world. As such, it was an important geopolitical space in the past. The immediate adoption of metallurgy may not have been revolutionary, but the aggregation of short and long-term impacts shows that the technology has had a big impact from value systems to subsistence and defence strategies to wealth accumulation and political power. As such, in order to understand society, it is important to study such influential technologies for they participated in societal continuity and change. Therefore, technology is fundamentally about humans, and if we learn about material and physical properties, we learn about the various aspects of humanity. This is the message that comes from Africa's preindustrial metallurgy as studied from an integrated view which develops a synergy between ethnographic, historical, geological and archaeological and archaeometallurgical points of view.

References

Alpern, S. B. (2005) Did they or didn't they invent it? Iron in sub-Saharan Africa. *History in Africa, 32,* 41–94.
Bachmann, H. (1982). *The Identification of slags from archaeological sites.* (Occasional Publication no 6). London: Institute of Archaeology.
Bhila, H. H. K. (1982). *Trade and politics in a Shona Kingdom: The Manyika and their African and Portuguese neighbours* (pp. 1575–1902). Harlow: Longman.
Brown, J. (1973). Early iron production. *Rhodesian Prehistory, 7,* 3–7.
Caton-Thompson, G. (1931). *The Zimbabwe culture: ruins and reactions.* London: F. Cass.
Celis, G., & Nzikobanyanka, E. (1976). *La métallurgie traditionnelle au Burundi: Techniques et croyances* (Archives d'anthropologie, no. 25). Tervuren: Musée royal de l'Afrique centrale.
Childs S. T. (1991). Style, technology and iron smelting furnaces in Bantu-speaking Africa. *Journal of Anthropological Archaeology, 10*(4), 332–59.
Chirikure, S. (2005). *Iron production in Iron Age Zimbabwe: Stagnation or innovation.* Unpublished doctoral dissertation, University College, London.
Chirikure, S. (2006). New light on Njanja iron working: Towards a systematic encounter between ethnohistory and archaeometallurgy. *South African Archaeological Bulletin, 61,* 142–151.
Chirikure, S. (2010). On evidence, ideas and fantasy: The origins of iron in sub-Saharan Africa. Thoughts on É. Zangato & AFC Holl's On the iron front. *Journal of AfricanArchaeology, 8*(1), 25–28.
Chirikure, S., & Bandama, F. (2014). Indigenous African furnace types and slag composition—Is there a Correlation? *Archaeometry, 56*(2), 296–312.
Chirikure, S., & Rehren, T. (2006). Iron smelting in pre-colonial Zimbabwe: Evidence for diachronic change from Swart Village and Baranda, Northern Zimbabwe. *Journal of African Archaeology, 4,* 37–54.
Cline, W. B. (1937). *Mining and metallurgy in Negro Africa* (No. 5). Menasha: George Banta Publishing Company.
Craddock, P. T. (2010). New paradigms for old iron: Thoughts on É. Zangato & A.F.C. Holl's on the Iron Front. *Journal of African Archaeology, 8,* 29–36.
Curtin, D., Feierman, S., Thompson, L., & Vansina, J. (1978). *African history.* London: Longman.
David, N., Heimann, R., Killick, D., & Wayman, M. (1989). Between bloomery and blast furnace: Mafa iron-smelting technology in North Cameroon. *African Archaeological Review, 7*(1), 183–208.
Goody, J. (1971). *Technology, tradition and the state in West Africa.* London: Oxford University.
Hall, M. 1987. *The changing past.* Cape Town: David Phillip.
Hansen, B. (1986). The complementarity of science and magic before the scientific revolution. *American Scientist, 74*(2), 128–136.
Heimann, R. B., Chirikure, S., & Killick, D. (2010). Mineralogical study of precolonial (1650–1850 CE) tin smelting slags from Rooiberg, Limpopo Province, South Africa. *European Journalof Mineralogy, 22*(5), 751–761.
Holl, A. (2000). Metals and precolonial African society. In M. Bisson, S. T. Childs, P. de Barros, & A. Holl (Eds.), *Ancient African Metallurgy: The sociocultural context* (pp. 1–81).Walnut Creek: AltaMira Press.
Holl, A. F. (2009). Early West African metallurgies: New data and old orthodoxy. *Journal of World Prehistory, 22*(4), 415–438.
Insoll, T. (2008). Negotiating the archaeology of destiny. An exploration of interpretive possibilities through Tallensi Shrines. *Journal of Social Archaeology, 8*(3), 380–403.

References

Kense, F. J., & Okoro, J. A. (1993). Changing perspectives on traditional iron production in West Africa. In T. Shaw, P. Sinclair, B. Andah, & A. Okpoko (Eds.), *The archaeology of Africa: Food, metals and towns* (pp. 449–458). London: Routledge.

Killick, D., & Fenn, T. (2012). Archaeometallurgy: The study of preindustrial mining and metallurgy. *Annual review of anthropology, 41,* 559–575.

Lechtman H. (1977). Style in technology—some early thoughts. In H. Lechtman & R. Merrill (Eds.), *Material culture: Styles, organization and dynamics of technology* (pp. 3–20). New York: West.

Miller, D., Killick, D., & Van der Merwe, N. J. (2001). Metal working in the Northern Lowveld, South Africa AD 1000–1890. *Journal of Field Archaeology, 28*(3–4), 3–4.

Mitchell, P. (2005). *African connections: Archaeological perspectives on Africa and the wider world.* Walnut Creek, CA: AltaMira.

Okafor, E. (1993). New evidence on early iron-smelting in Southeastern Nigeria. In T. Shaw, P. Sinclair, B. Andah, & A. Okpoko (Eds.), *The archaeology of Africa: Food, metals, and towns* (pp. 432–448). London: Routledge.

Plug, I., & Pistorius, J. C. C. (1999). Animal remains from industrial Iron Age communities in Phalaborwa, South Africa. *African Archaeological Review, 16*(3), 155–184.

Prendergast, M. D. (1975). A new furnace type from the Darwendale Dam basin. *Rhodesian Prehistory, 7*(14), 16–20.

Prestholdt, J. (2008). *Domesticating the world: African consumerism and the genealogies of globalization.* Berkeley: University of California Press.

Prestholdt, J. (2004). On the global repercussions of East African consumerism. *The American Historical Review, 109*(3), 755–781.

Pryce, T. O., Pigott, V. C., Martinón-Torres, M., & Rehren, T. (2010). Prehistoric copper production and technological reproduction in the Khao Wong Prachan Valley of Central Thailand. *Archaeological and Anthropological Sciences, 2*(4), 237–264.

Rickard, T. A. (1939). The primitive smelting of iron. *American Journal of Archaeology, 43*(1), 85–101.

Robertshaw, P. (Ed.). (1990). *A history of African archaeology.* London: James Curry.

Rodney, W. (1974). *How Europe underdeveloped Africa.* Dar es Salam: Dar es Salam University Press.

Schmidt, P. R. (1978). *Historical Archaeology: A structural approach in an African culture.* Westport: Greenwood Press.

Schmidt, P. R. (1997). *Iron technology in East Africa: Symbolism, science, and archaeology.* Bloomington: Indiana University Press.

Schmidt, P. R. (2001). Resisting homogenisation and recovering variation and innovation in African iron smelting. *Mediterranean Archaeology, 14,* 219–27.

Schmidt, P. R. (2009). Tropes, materiality, and ritual embodiment of African iron smelting furnaces as human figures. *Journal of Archaeological Method and Theory, 16*(3), 262–282.

Smith, C. S. (1981). *A search for structure. Selected essays on science, art and history.* Cambridge: MIT Press.

Stahl, A. B. (2014a). Africa in the World: (Re)centering African history through archaeology. *Journal of Anthropological Research, 70*(1), 5–33.

Stahl, A. B. (2014b). Metal working and ritualization: Negotiating change through improvisational practice in Banda, Ghana, AD 1300–1650. In L. Overholtzer & C. Robin (Eds.), *The materiality of everyday life. Archaeology papers of the American Anthropological society.* Arlington: American Anthropological Society.

Sutton, J. 1985. Temporal and spatial variability in African iron furnace. In R. Haaland & P. L. Shinnie (Eds.), *African Iron working-ancient and traditional* (pp. 164–197). Oslo: Norwegian University Press.

Van Warmelo, N. J. 1935. *A preliminary survey of the Bantu tribes of South Africa.* Pretoria: Government Printer.

Yacovone, D. (2002). Editor's Introduction. Special issue, race & slavery. *Massachusetts Historical Review, 4,* i–iii.

Index

A
Agriculture, 43, 126, 127, 133
    swidden, 126
Akan, 42, 108, 110
Akjoujt, 20, 24, 29, 78, 105, 152
Amos G. Beman, 159
Annealing, 103, 115
Anthropology, 2, 10–12, 63, 88, 156
    of mining, 2, 12, 35, 38, 40, 43, 44, 46, 49, 51, 56, 64
    of smelting, 63, 64, 67, 75, 79, 88,
Anthropology of metal fabrication, 119
Archaeology, 3, 5, 6, 7, 10–12, 43, 49, 63, 65, 78, 85, 134, 154–159
    of mining, 43
    of smelting, 43
    of smithing, 110
Asante, 29, 115, 121, 127, 131

B
Backfilling, 47, 56
Backward Africa, 153
Bag, 67
Beliefs, 37, 40, 55, 88, 93, 120, 134, 154
Bellows, 61, 63, 67, 70, 72, 75, 81, 84, 88, 92, 102, 103
    goatskin, 106, 111, 112
Bowl, 67
    ceramic, 70
    furnaces, 65, 70, 80, 84, 85, 155, 156
Bridging conceptual boundaries, 151
Buhaya, 5, 27, 83, 84, 91
Buhen, 19, 71
Burkina Faso, 106

C
Copper ingots, 70, 112, 118, 139
Cowrie, 142

D
Decorated furnaces, 90
Dekpassanware, 64, 75, 107
Diffusion, 50, 152
Dimi of Ethiopia, 111

E
Economy, 127, 157
El Tio, 37, 55
Ethnography of, 87
    mining, 109
    smelting, 104, 105, 109, 110, 184
    smithing, 16
Exotic imports, 144, 145
Expressive objects, 99

F
Fire setting, 44, 49
Fitola, 26, 27
Furnace types, 7, 24, 27, 63, 67, 76, 79, 80, 153, 160

G
Glass beads, 25, 116, 127, 136, 137, 141–144, 146, 152, 158, 159
Gold weights, 108
Gold working, 115–117, 121
Great Zimbabwe, 12, 30, 38, 86, 116, 117, 119, 120, 127, 128, 131, 134, 136, 137, 139

H
Hammering, 101–104, 106, 107, 111, 116
Hoisting, 50, 53
    ore, 46
Human reproduction, 88

I
Igbo Ukwu, 29, 78, 108, 127, 143
Imported materials, 106, 116, 143, 144
Improvisation, 88, 90
Independent origins, 20, 27, 120
Iron and kingship, 134

K
Kansanshi, 45, 65, 81, 82
Kaonde group, 44, 64
Khami, 116, 119, 127, 131, 136

L
Leija, 26, 27
Lobola, 120
Low shaft, 65, 67, 78, 80, 84, 155, 156
Luminescence dates, 27

M
Mabveni, 23
Magic, 6, 10, 53, 55, 61, 88, 154, 161
Magnetite, 6, 8, 37, 41, 43, 52, 64, 77, 79, 87
    sands, 41, 43, 64
    skins, 77
Malachite, 52, 64, 70, 71, 81
Mandara Mountains, 6, 41, 107, 140, 160
Mapela, 131, 134, 137
Mapungubwe, 116, 118, 120, 127, 131, 134, 136, 137
Mastaba of Mereruka, 71
Meroe, 19, 27, 39, 67, 72, 75, 92, 121, 126, 140
Metals and socio-political complexity, 130, 131, 133, 145
Metals and urbanism, 130, 131
Metaphors, 11, 79, 89
    of reproduction, 88, 90. 92
    socio-cultural, 88
Mining tools, 49, 53
Molds, 81, 103, 104, 108, 112, 116, 117
    soapstone, 117
    steatite, 108
Musuku, 114, 115, 138

N
Natural draught, 65–67, 75, 76, 80, 82, 86, 106, 107, 155, 156, 160
Nubian Pharaohs, 21

O
Obui, 25
Old wood problem, 24, 28
Open mining, 40, 42, 44–46, 49, 50
Ores
    copper, 21, 37, 38, 40, 46, 125, 157

P
Pictographs in Egyptian tombs, 70
Political power, 134–136, 158, 161
Pot, 67, 70, 72, 79, 84, 92, 108

R
Radiocarbon black hole, 17, 29, 77, 105, 151
Reproduction of society, 161
Ritual and magic, 6
Rwiyange, 23

S
Sanga, 181, 130
Scale of production
    fabrication, 153
    mining, 12
    smelting, 39, 76
Shackles, 107, 108, 140
Shona, 5, 6, 42, 45, 51, 52, 64, 120, 136, 138, 140, 141
Smelting
    iron, 152
Smelting East Africa, 92, 152
Smelting Egypt, Nubia and North Africa, 63
Smelting Southern Africa, 84, 85, 87, 88, 92
Smelting West Africa, 27, 64, 74
Smithing Egypt, Nubia and North Africa, 63
Southern Zambezia, 5, 51, 52, 112, 139
Sukur, 106, 107, 121, 140
Swahili, 12, 30, 115, 141

T
Technology transfer, 19, 29, 138, 140, 144, 152, 157, 161
Trade and exchange, 7, 121, 126, 127, 136, 138, 160

U
Underground mining, 37, 38, 40, 45–47, 49, 50, 56
Urbanism, 131

V
Value transfer, 142, 152

W
Wadi Dara, 64, 69
Waldadé, 77
Wandala, 130, 140
Wire drawing, 106, 117, 118, 140

Y
Yeke, 44, 81

The manufacturer's authorised representative in the EU is Springer Nature Customer Service Centre GmbH, Europaplatz 3, 69115 Heidelberg, Germany. If you have any concerns regarding our products, please contact ProductSafety@springernature.com

Printed and bound by CPI Group (UK) Ltd, Croydon, CR0 4YY

23/03/2026

02076665-0003